Unsettled Borders

DISSIDENT ACTS
Edited by Diana Taylor and Macarena Gómez-Barris

Unsettled Borders

The Militarized Science

of Surveillance on Sacred

Indigenous Land

FELICITY AMAYA SCHAEFFER

DUKE UNIVERSITY PRESS
Durham and London   2022

Project editor: Lisa Lawley
Designed by Courtney Leigh Richardson
Typeset in Quadraat Sans and Portrait by
Westchester Publishing Services

Library of Congress Cataloging-in-Publication Data
Names: Schaeffer, Felicity Amaya. author.
Title: Unsettled borders : the militarized science of surveillance on
sacred indigenous land / Felicity Amaya Schaeffer.
Other titles: Dissident acts.
Description: Durham : Duke University Press, 2022. | Series: Dissident
acts | Includes bibliographical references and index.
Identifiers: LCCN 2021031681 (print)
LCCN 2021031682 (ebook)
ISBN 9781478015321 (hardcover)
ISBN 9781478017943 (paperback)
ISBN 9781478022565 (ebook)
Subjects: LCSH: Border crossing—Social aspects—Mexican-American
Border Region. | Border security—Mexican-American Border Region. |
Indigenous peoples—Mexican-American Border Region—Social
conditions. | Traditional ecological knowledge—Mexican-American
Border Region. | Indian activists—Mexican-American Border Region. |
Mexican-American Border Region—Emigration and immigration—
Social aspects. | BISAC: SOCIAL SCIENCE / Emigration & Immigration. |
SOCIAL SCIENCE / Indigenous Studies | SCIENCE / Philosophy &
Social Aspects
Classification: LCC JV6475 .S525 2022 (print) | LCC JV6475 (ebook) |
DDC 304.8/721—dc23/eng/20220118
LC record available at https://lccn.loc.gov/2021031681
LC ebook record available at https://lccn.loc.gov/2021031682

Organ Pipe Cactus National Monument, Ajo, Arizona, February 2020.
Construction of former US President Donald Trump's border wall in
the area has destroyed many species, including ancient saguaro cacti.
© Sandy Huffaker. Courtesy Getty Images/afp, and the artist.

I dedicate this book to my late father, whose quiet ways
could move mountains. . . .
And to my mom, whose heart and soul feeds the mountain. . . .

CONTENTS

PREFACE *From TimeSpaces of Dispossession*
to the Forging of Indigenous Relations with Land

I began this book out of a deep concern for how the border has become normalized through a militarized approach to security and especially how surveillance technologies have enabled the state to "see" deception and threat on brown bodies alienated from the contexts that send them fleeing untenable lives back home. Each story of migrant death, rape, detention, deportation, and hardship awakened an aching memory of my own family history of migration from Mexico to the United States. My grandparents were forced to leave everything behind in the small town of Bustamante, located in northern Mexico, after the 1910 revolution brought widespread armed conflict across the northern territories of Mexico. Fleeing displacement from their land, they traveled with thousands of others about fifty miles to *el norte*, on the US side of the Texas border.

On this journey of crossing from Mexico to the United States—despite not speaking English and not being equipped with many resources—they tried, or we might say were coerced, to become white (although this mostly failed). As part of a long history of legal and thus social inclusion for Mexicans who assimilated into whiteness, my grandparents worked hard and declared themselves *Spanish* Mexicans to distinguish themselves from the stereotype of the rebellious, or lazy, *indio* in Mexico and from the stigma of the American Indian primitive in the United States.[1] As part of this ascent into whiteness, my grandparents never discussed the hardships of this journey with their children. They worked in the fields, then opened small businesses, and never looked back. By the time their daughter (my mother) fled the racism in Texas (especially toward her and a few of her nine other siblings who had darker skin) and moved to California, her Spanish was broken, maimed by a dominant culture that would not tolerate difference. Like her own mother, she left behind these scattered stories of the past in Texas, herself a refugee from a homeland that held two generations of unspoken loss.

It should be no surprise that when stories of the past are severed and memories lie dormant like bones barely jutting out of sandy desert, illness and melancholic loss can take hold. This happened to some in my family. When much younger, my cousin and I heard the elder women in my family mention in Spanish, under their breath, that we had the "gift," but they never explained what that meant. It wasn't clear whether these gifts were a curse or an enviable skill. From these generational silences, I took refuge in Gloria Anzaldúa's *Borderlands/La Frontera: The New Mestiza*, where I finally found expression for a pain I tenderly held in my bones, the sense of being outcast by dominant stories and histories that prevented me from calling any place home, but also the skill that haunted those outcast: that of seeing phenomena not readily apparent through Western ways of seeing.

Anzaldúa's theorization of the borderlands held the wounds of the displaced but also the potential for digging deeper into a past that was not simply lost but suppressed by dominant patriarchal knowledges that had the power to define the world.[2] Against these erasures, Anzaldúa unpacked inherited knowledges shaping cultural expressions to invent her own kin relations with matriarchal Indigenous cosmologies, knowledges suppressed by the patriarchal culture of Aztlán, the supposed homeland of Chicanos who descended from the Aztecs.[3] As a child, she invented and remembered stories, visions, and dreams of the surreal and intimate encounters with nonhuman kin, especially animal beings (such as serpents, owls, and deer) with whom she could converse, sometimes without the use of language. With the aid of sci-fi literature, meditation, ceremony, and inner work, she traversed the world with a range of lively beings who, like Western technologies, shed light on her journey to expand consciousness, vision, and ways of knowing and being that were demonized as primitive, unscientific, magical, superstitious, and idolatrous. While understandably critiqued for making claims to an Indigenous Mesoamerican past, rather than engaging with Native peoples today, Anzaldúa's poetic writing daringly ruptured the limits of the Western epistemologies that prop up the national timespace. This opening of alternative possible worlds inspired my search to decolonize the borderlands through and beyond militarized surveillance technologies.

The decolonial turn in transnational Xicana and Latinx studies is varied and for many is inseparable from a queer methodology oriented to seeing beyond Western heteronormative and masculinist perspectives.[4] This desire to reclaim an Indigenous heritage wiped out by Spanish colonialism, Mexican nationalism, and US empire is understandable, and yet the desire for belonging and access to land-based claims to the Southwest is much more complex when we center the settler history of this same land, which was stolen from Native Americans whose own claims to land are trampled.

When Anzaldúa said the land was, is, and will again be Indigenous, she meant two things at the same time, that the land would be returned to the Indigenous peoples on both sides of the border, including mestizas or Xicanas who would reclaim their ties to land by learning the ways of those native to the land. While she inadvertently obscured the long presence of Native peoples across the borderlands and relegated Indigeneity to the past, there is a glimmer of possible worldings we may want to hold onto. Anzaldúa's *Borderlands* also holds out an offering, or invitation, to return the land to those who practice Indigenous relations with land. She refused to live severed from the past and entreated us to learn with Mesoamerican cosmologies (however much she treated these knowledges as universal)—or perhaps she'd say today with the Native peoples who continue to live on and make claims to the land—that many of us have been raised not to see and know. She followed Aztec and Mesoamerican footprints as a journey back to heal the consequences of being ruptured from this past that was also tied to land. While she had no ties to a Indígena tribe in Mexico, she—like so many of us—sought knowledges and cultural practices to help her understand these unexplainable gifts of insight, visions, dreams, and a spiritual consciousness that offers wisdom from times before we were born and long after our human life on earth has ended. I am grateful Xicana scholars have named this search for connection to our Indígena past as a de-tribalized or de-Indigenized people.[5] The possibility for decolonizing the multiple timespace ruptures with land across the world (stolen by genocide, enslavement, occupation, boarding schools, citizenship, rights, law, gentrification, capitalism, and so forth) might fruitfully take us back to where women of color and third world women began organizing in the 1970s and 1980s, by honoring our shared stories of displacement as a start to remembering, learning, and building a pluriverse of possible land-based Xicana Indigeneities. These alliances are possible only when the land is returned to the Native peoples of these lands.

ACKNOWLEDGMENTS

There were many beginnings and sojourns that finally took root thanks to so many who have moved my thinking along during the writing of this book. The seed for this project began with a paper I presented at the Gloria Anzaldúa Conference that Cindy Cruz and I co-organized in 2015, called "The Feminist Architecture of Gloria E. Anzaldúa: New Translations, Crossings, and Pedagogies in Anzaldúan Thought." I'd like to thank the many attendees who encouraged me to continue with this work: Aida Hurtado, Rosa-Linda Fregoso, Marcia Ochoa, Bettina Aptheker, María Lugones (who joins us now in spirit), Pedro DiPietro, Laura Pérez, Wanda Alarcon, and many others.

I am heartened to be surrounded by so many creative thinkers here at the University of California, Santa Cruz, in the Feminist Studies Department, who always push me to think beyond the bounds. I am indebted to Neda Atanasoski, Marcia Ochoa, Nick Mitchell, Neel Ahuja, Bettina Aptheker, Jenny Kelly, and, more recently, Katie Keliiaa for their friendship, engagement, and thought-provoking conversations with me. I am also grateful for the many years of intellectual support from my Latin American and Latino Studies colleagues and comrades, Catherine Ramirez and Sylvanna Falcón, who fostered inspiring dialogues while directing the Chicana/Latina Research Center and now the Research Center of the Americas, both centers that have generously funded the research for this book. A warm thanks to many others for their friendship, support, and *pláticas* (conversations): Cecelia Rivas, Pat Zavella, Gabriella Arredondo, John Jota Leaños, Julietta Hua, Grace Hong, Kalindi Vora, Kasturi Ray, Alicia Schmidt-Camacho, Brian Klopotek, Bianet Castellanos, Joanna Barker, and Mishuana Goeman.

A warm thanks for my powerful writing sessions with Beth Haas, whose sharp critical insight has pushed me to be a better thinker and writer. And to Edhi Shanken for his heroic editing at the final hour. A shout-out to those of you whose work and friendship hold me up from afar: Ines Dolores Casillas, Deb

Vargas, Maylei Blackwell, Susy Zepeda, and Macarena Gómez-Barris, whom I want to thank in particular for being such an ardent advocate of this book.

My institution, the University of California, Santa Cruz, provided much support for this book, including a grant from the Fellows Academy in 2019–20, a writing program initiated by former executive vice chancellor Marlene Tromp. Through this inaugural writing program, I was lucky to have a sharp writing group with Stacy L. Kamehiro, Catherine Ramirez, and Megan Thomas. And a special thanks to Stacy for spearheading such a relaxing writing retreat together that helped move this project to the finish line. I am also grateful that my research in the Yucatán of Mexico was supported by a faculty research grant awarded by the Committee on Research from the University of California, Santa Cruz.

Early versions of this project benefited from the collaborative insight of participants of Illegality Regimes, a conference in 2013 organized by Juan Amaya Castro and Bas Schotel at Vrije Universiteit Amsterdam and the workshop The Process of Imaging/The Imaging of Processes, at the Seventh Annual New Materialism Conference in Warsaw, Poland, in 2016. And thanks to George Lipsitz for supporting the publication of an early version of one of the chapters and to my co-collaborators on the Sawyer-Mellon grant on Non-Citizenship funded by the Andrew W. Mellon Foundation: Sylvanna Falcón, Steve McKay, Juan Poblete, and especially again Catherine Ramirez. A special thanks to Miranda Outman, who was my amazing editor of "Spirit Matters," a paper published by the journal *Signs: Journal of Women in Culture and Society*.

I am especially thankful to the blind reviewers at Duke University Press, whose feedback made a lasting imprint on the book, and to Gisela Fosado and Alejandra Mejia for keeping me on track.

I thank my amazing posse of graduate students whose research inspires me: Claire Urbanski, Dana Ahern, Victoria Sanchez, Yvonne Sherwood, Natalie Gonzales, Alfredo Reyes, Vivian Underhill, Ryan King, and Cecelia Lie. Much gratitude to Claire Urbanski, who aided me with research and careful editing at a crucial time, and to Dana Ahern, whom I was lucky to have as a research assistant at an earlier stage in the project.

I also want to acknowledge those who guided and accompanied me in doing the deep work of bringing my Xicana political identity closer to Indigenous politics with the land I live on (the unceded territory of the Awaswas-speaking Uypi tribe): the Amuh Mutsun tribal band working to heal and restore this land, the Oholone, and the Coastanoans of Indian Country, especially Ann Marie Sayers, who generously invites us to their land for the Women's Encuentros. I am thankful for the women holding this sacred space for healing and community:

Reyna, Tia Rocky, Susy, and Iriany, and so many others. And to Iriany in particular for coming into our home and life and inviting me into the powerful ways
of ceremony, sweat lodges, and songs. I am also deeply grateful to my O'odham
friends for inviting me to their homeland and for taking such good care of
Iriany and me while we were there: Mike, Cesca, Troy, and Ofelia. And to Fred,
for his songs and blessings, whose spirit now flies like the eagle protecting his
people/land.

And, finally, I want to thank my extended family: Amaya and Diego, whom
I am in awe of at every turn. Their deep insight and hormonal swings keep me
on my toes, ground and unravel me (in the best ways), and are also the source of
incredible joy, wonder, and lightness. To Neda and Jeanette for accompanying
me on a research trip tattooed on my heart. To my mother, whose love and support has carried me far along on this journey. And to my brother, who always
brings me back to myself when I become unmoored.

Unsettled Borders theorizes the contested status of today's militarized security technologies deployed at the US-Mexico border (and beyond) from the perspective of Native Americans who inhabit these lands. I use the term *unsettled* to foreground the contestations of the Apache, Tohono O'odham, and Yucatec Maya against the settler state's ongoing attempts to dispossess them of their ancestral lands, to border or contain their movement, and to eradicate their sacred relationality with land in the name of securing the nation.[1] As the escalating financial and human costs of securing the US border reach obscene levels, the Arizona-Mexico border, in particular, has become a profitable laboratory for the most cutting-edge innovations in automated surveillance. Despite these costly and violent developments, little is known about the rise of militarized surveillance along the border and the historical conditions, knowledges, and desires that undergird its technological eye.

In the process of tracking how the US border in Arizona (and borders around the world) has become militarized, many questions arose about how automated surveillance technologies allowed the state to absolve itself of violence and to extend its reach beyond US national boundaries, law, and jurisdiction, especially onto Native American reservations. The increasing militarization of the border demonstrates the ramping up of US control over the sovereignty of people and land.[2] Not only do the surveillance sensors used along borders today delve deeper into the body, across Native reservations, high up in the air, and beneath the ground, but never before has the government circumvented so many of its own laws to build and govern the virtual and physical border wall. This begged the question, How did we get to this point? And I realized that until we understand the logic driving the development of surveillance at borders, we may not find the best solutions for imagining and realizing a way out of this sprawling gaze of the state. Yet as I followed the military research and development driving surveillance technologies along the Arizona-O'odham-Mexico border farther back, another story emerged that significantly challenged and expanded my personal, academic, and political commitments.

I have spent most of my academic career preoccupied with Xicana/Latinx and Latin American immigration and decolonial border issues. It was not until undertaking this project that I came to recognize a glaring absence in these fields: with few exceptions they failed to consider the struggles of Native Americans against the state's occupation of their land.[3] This oversight is stunning when we consider that today there are twenty-six US tribal nations recognized by the US federal government living in the US-Mexico border region and a smaller number of Indigenous peoples whose land and people cross into Mexico—including the Tohono O'odham, the Yaqui (Yoeme), the Cocopah, the Kumeyaay, the Pai, the Apaches, the Tigua (Tiwa), and the Kickapoo.[4] To come to terms with this widespread present-absence that contributes to the relegation of Native peoples to the past, the story of the border must be centered in the ongoing settler colonial dispossession of Indigenous peoples *and* migrant-refugees.[5]

I tell this story from Fort Huachuca, a military fort in Arizona close to the border (the ancestral land of various groups of Apache, some which seasonally return), from the Tohono O'odham reservation, and even from the Yucatán, where Maya women are fighting to preserve their land. Not only do Native peoples live on both sides of the border, but many migrants crossing this land are displaced Indigenous peoples, including Maya from southern Mexico, Guatemala, and Honduras.

The fields in which the book intervenes—Latinx immigration and border studies as well as feminist science and technology studies—rarely engage, and if so, only recently, in debates in Indigenous studies.[6] To bridge these divides, I not only draw from Indigenous studies scholars' critical engagements with borders, empire, technology, and land-based epistemologies, but I also build on lessons learned from Native American and Xicana elders, and gained insights from Maya/bees, O'odham/saguaros, and Apache/mountains. From these knowledges I theorize the concept of sacredscience. Sacredscience considers Native American understandings with land as *scientific*, a methodology of collective intelligence that regenerates ancestral knowledges to sustain sacred intrarelationality with land. This term is elaborated in the final section of the introduction and runs throughout the book.

The research for this book and the recent expansion of border and migration studies into Indigenous epistemologies radically jolted my understanding of the borderlands. Unquestioned perspectives about migrants' experiences in Latinx studies must be rethought when it participates in the erasure of Native

Americans from history *and*, especially, from the ongoing violence at the border. Given that scholarship, media, and films on immigration and borders focus on the brutal journey of border-crossers, we have failed to also account for the challenges faced by Native Americans such as the Tohono O'odham, who are under US occupation by the border-security-industrial complex on their own land. They, too, face a carceral logic of containment, detention, prisons, and even deportation. In addition, focusing on migrants' rights to documentation without considering Native American sovereignty, and on assimilation without discussing policies aimed at Native extinction, renders unthinkable the land-based struggles of the O'odham, Maya, and Apache against the theft of unceded territories along the border and beyond that hold their ancestral knowledges. Even the pro-immigrant political outcry—the United States is a nation of immigrants!—diminishes Native American ancestral claims to land.[7]

Along with Jodi Byrd, Roxanne Dunbar-Ortiz, and María Josefina Saldaña-Portillo, I consider the ways the empire of border regions are "produced through colonial encounters with Indigeneity."[8] This more recent reorienting of Latinx/Latin American/border studies through Indigenous studies has usefully challenged intellectual ruts and expanded political collaborations forged by activists and scholars including Maylei Blackwell, M. Bianet Castellaños, Macarena Gómez-Barris, Audra Simpson, and many others.[9] The absenting of Native Americans and the history of the Apache Wars (1861-1900) as formative in the emergence of militarized border surveillance today warrants more careful and sustained attention if we are to build a broad-based politics aimed at abolishing the carceral and imperial logics of physical and virtual borders as well as border epistemologies.

Unsettled Borders follows the technoscientific logics of settlement that reproduce colonial binaries of civilized versus primitive land and bodies, which have consequences for how borders are seen, or imagined, and thus governed, as well as how they can be broken down. The book traces the militarization of the US-Mexico border, and especially today's automated border surveillance, to Fort Huachuca, a high-tech military intelligence hub twenty miles from the Arizona-Mexico border. The fort was first erected in 1877 as a military garrison, tasked to track and control fugitive movements by "marauding Indians."[10] The most notable of these fugitives were the Chiricahua Apache, who refused to be moved onto reservations, who continued to raid encroaching settlers, and who cleverly escaped Euro-American cavalry troops by fleeing to Mexico. Many of the tools deployed at the border today—drones, surveillance gadgetry, aerial balloons, and other military techniques—are developed and tested at Fort Huachuca.

From the historical accounts of Fort Huachuca, the Apache Wars emerge as the incubator for innovations in border control.[11] At this fort, surrounded by the Huachuca mountain range considered sacred to the Apache and other Native peoples in the area, the science of military innovation began, and the Indian Wars supposedly came to a close.[12] According to Fort Huachuca's museums and archival documents, the Indian scout was the most significant innovation in the cavalry's arsenal. Unable to capture Apaches who easily maneuvered across the "wild" and unknown terrain of the western frontier and into Mexico, lieutenants hired Indian scouts to join the front lines of battle, significantly extending the army's ability to see and move across rugged mountainous terrain and impenetrable desert. As an early military technology, the visual acuity of Indian scouts—what I call Nativision—was abstracted by the US military as the "eyes of the army," converting Native bodies and cognitive practices into biotech sensors, information conduits, and cryptography that extended the state's vision, communication, and control over territories. However effective Indian scouts were in aiding the army to track Apache resisters, Native knowledges were significantly distorted in the process of being translated into Western technoscientific discovery and innovations. Once converted into technologies, the source of that knowledge disappeared from the memory of the settler nation. The military's extraction of Nativision into its innovative arsenal of control, converts colonial genocide into a technoevolutionary story. This story naturalizes and authorizes the US nation's technological emergence from the savage past into a civilized futurity.

I argue that the colonial construct of the Indian "savage" is the original threat justifying militarized approaches to border security. I further argue that Chiricahua Apache and other Native American warrior intelligence inspires the military innovations that weaponize border surveillance around the globe today.[13] In other words, not only are the Indian Wars an overlooked moment in most historical treatments of the rise of the US-Mexico border, but Indigeneity and Indian Country are figurations of primitive otherness that bolster the empire of border security domestically and around the world.[14] For example, in chapter 1, I demonstrate how military innovations in border security, settlement, and dispossession gleaned from Apache warriors and used to settle the frontier during the late 1870s were deployed less than a decade later in the Philippines and Puerto Rico. And as I show in chapter 3, bordering technologies used against Palestinians in Israel, are deployed against the O'odham on the Arizona border, while drones and surveillance developed at Fort Huachuca are repurposed in Iraq and Afghanistan.

At Fort Huachuca, the inauguration of military surveillance cannibalizes Native knowledge into the military's own embodied arsenal of supposedly automated vision, an act that erases the Apache's sophisticated relational knowledge with land that threatened the settlement of the West. These place-based worldviews are converted into armaments of military security and then used to dispossess the very same peoples, to disarm them of the collectively held land that has long sustained their freedom to live according to the rhythms of their land. Fighting off threats at the border continues to be imagined as a war between the primitive and the modern, programmed into the scientific methods and technological innovations that are key to maintaining territorial control. This belief justifies the constant monitoring and control of bodies and borders while reifying the primitive/modern hierarchy.

Early military intelligence—influenced by colonial evolutionary science—converted Native American sacredsciences into a resource, or data to be extracted. Thus, Indian (time) and Indian Country (space) are coded into the militarized optics of border surveillance. The rise of scientific objectivity masks the context of its emergence, such as the colonial theories of evolution undergirding the science of comparison (anthropology, ethnology, ethnography, physiognomy, and archaeology). Thus, I trace a trajectory from the technoscientific knowledges that buttressed territorial borders during the Apache Wars (1848-1924) to the automated intelligence surveillance programmed into today's border wars (ground and aerial surveillance, remote communication technologies, data sharing, computational biology and the algorithmic processing of big data). Often missed is how these technoscientific knowledges perpetuate racial violence *and* Native dispossession, eradicating sacred ties to land.

Indians were romanticized as fiercely independent and thus ideal subjects exemplifying a rebellious individualism antagonistic to industrialized docility but were also considered a dangerous impediment to manifest destiny, and thus to settler wealth and security. As argued by Philip Deloria, American fondness of "playing Indian," or nurturing the "wild Indian within" has a long and complex history.[15] Indigeneity has long inspired the spirit of Americanness (shaping the Constitution, settler masculinity, and military strategy), thus anxiously situating Native peoples as both inside and outside American identity, law, and land.[16] For these reasons, the territorial proximity of Indigeneity within the American psyche and nation was and is of concern to the status quo and must be relegated to the past or contained within or isolated from civilized nations.

In fact, Native skills were embraced as patriotic when used in the service of US empire and as dangerous when they stymied Western progress. Military intelligence has long been fascinated with the secrets of Apache warriors' ability to communicate across vast distances. This colonial imaginary continues to drive military strategies. During World Wars I and II, the army capitalized on Choctaw and then Navajo code (or wind) talkers, who used their language as code to transmit secret messages across the airwaves. Then, starting in the early 1970s and continuing today, the Border Patrol has relied on the superior tracking skills of Native Americans to detect the elusive presence of migrant footsteps, forming an all-Native Border Patrol on the Tohono O'odham reservation called the Shadow Wolves. More recently, a broad-based group of Native American border patrol called NATIVE Immigration and Customs Enforcement sends the Shadow Wolves, imagined as fearless hunters of animals and humans, to borders around the world owing to their mythic reputation in the media as "manhunters."[17]

Tracking Footprints: Remapping the Borderlands

Contesting the military's turn to Native tracking and surveillance as technologies of border control, *Unsettled Borders* remaps the border through a focus on the O'odham, Maya, and Apache cosmologies of footprints. To follow footprints entails an understanding of place undertaken through the study of material and ephemeral ancestral imprints left on the land as well as storied relation to place. For instance, the Huachuca mountain range is an ancestor; a sacred portal linking the past, present, and future-to-come; and an elevated being that offers prophetic visions of other times and places, privileged knowledge that could influence decisions made in the present. One important Apache vision originating from these mountains told of the coming of a strange people with light hair and blue eyes that some interpreted as the coming of death, perched atop large dogs with pointy ears (what they later learned to be white men on horses). Starry-night ceremonial visions warned the Chiricahua of the arrival of the Spaniards and later Euro-American settlers before anyone could see them arriving from the mountaintop. Seeing footsteps was thus a complex activity that included prophecies; ancient memory; visions; dreams; the deep study of stars, land and animals; and elevated viewpoints that materialized a sacred connection with place.

Before the advent of sixteenth-century Spanish maps in Mexico, Maya maps illustrated footprints across land, marking sacred places alongside prominent animals, plants, and places that supported and defined the critical beings that support human flourishing. Mapping land in this way communicated the character of land through a recognition of important inhabitants and sites, a living and fluid place in which a segregated border was unthinkable.[18] And for the O'odham and Apache today, movement across land to hunt, to harvest, and to engage in pilgrimages connects them to a phenomenological orientation of belonging with ancestors who made similar journeys. Since the body-as-land holds memory, one can think of footprints as a praxis of awakening collective sensual bonds with land. To follow the ever-shifting footprints of animals, invaders, clouds, stars, plants, rocks, and winds is to learn from, and with, diverse beings rather than to map, or to contain life into knowable (dead) objects that can be converted into property.

These embodied knowledges with the more-than-human world hold insurgent knowledges, memories of ancestral presence and ways of belonging to and with land. Rather than reproducing state definitions of membership or citizenship—through blood quantum, biometric documents, or DNA—orienting oneself to ancestral paths imprints the land with footprints that define belonging as a collective relation with the beings of a particular place.[19] The past is a well-worn path or trail, a footprint, that is *alive* as a companion that guides one's actions in the present-future. In other words, a footprint is not simply a sign of the past imprinted on earth but a respected ancestral guide that collectively orients how Native Americans care for and preserve all the animal and plant nations and their "worldings" across time and space. In this book, ancestral practices with more-than-human beings—mountains, wind, bees, water, and saguaros—serve as guides who preserve vital practices of intrabecoming that sustain people and their land. When one fails to follow these footprints, they may go extinct, along with the people.

Orienting oneself to the pathways of ancestral footprints draws inspiration from Indigenous land-based knowledge practices as well as postcolonial queer studies scholars such as Sara Ahmed. Ahmed argues that to be oriented in space is to follow footprints toward a path "as a trace of past journeys," an alignment with others that sustains communal bonds.[20] By following footprints we see an incommensurate claim to the same regions of the borderlands that bring Latinx and Maya migrants onto the ancestral lands of Native Americans.[21] With these incommensurate relations to the borderland in mind, I see possibility for solidarity when we first orient our understanding of the border from the autonomous perspective of Native peoples, or the Indigenous borderlands. There are numerous possibilities for being in alignment with Native Americans' land claims, especially when we orient ourselves to the intimate ties that entangle a people with a particular place. For this reason, I turn to decolonial perceptual cosmologies that materialize sensual embodied relation to place through Indigenous, Xicana, and queer methods and practices such as Gloria Anzaldúa's poetic entanglement with land. In her pathbreaking book *Borderlands/La Frontera: The New Mestiza*, I consider Anzaldúa's less theorized Xicana sensual methodology with the more-than-human evident in the poetic ruptures of her imaginal writing. While she failed to address the actual histories and knowledges of Native tribes along the border today, there is much to glean from her training as a *naguala* or *chamana*: one who perceives beyond the borders of the self and other. Through a sensual engagement with the more-than-human that exceeds the temporal and spatial confines of the borders and their ontoepistemological limits, Anzaldúa's praxis is politically engaged with a return to Indigenous relation with land in Mexico and the United States.[22]

I also follow the footprints of Indigenous presence and absence by tracing how military surveillance technologies, what I call *settler surveillance*, void the land of all perspectives except one—that of the state. Automated state seeing empties the land of Native presence, even as it scans landscapes for Native threat—ironically, by deploying technologies derived from Indian scouts. As a tool of alienation, settler surveillance abstracts relational ties with land through an extractive relation to phenomena. This extractive vision has long shaped the scientific gaze, a way of seeing that estranges, abstracts, and extracts as its mode of seeing and knowing. *Unsettled Borders* traces the flashpoints of Native presence foundational to the military innovation and design of surveillance. At the same time, each chapter unsettles technoscientific borders by questioning how Enlightenment science and technologies desacralize land and bodies by extracting and automating Nativision into Western ways of knowing and seeing. Instead of segregating modern technology and science from Indigenous traditional ecological knowledge (TEK), the book proposes the term *sacredscience* to confuse the temporal fabrications and historical erasures that segregate tradition from science, the human from the nonhuman, the subject from the object, the local from the universal, the past from the future, and so on.

By attending to a range of struggles, demands, and knowledge practices shared with me, I have come to see how entangled many of us are in settler ways of seeing and unseeing that rob Native peoples (and all of us!) of knowledges with land critical to their/our healing and sustenance with land, now and long into the future. There is much to learn from Indigenous feminist, Latinx, and queer sensing, feeling, and seeing that hone in on sensual ways of relating with the land and each other that crowd out the destructive vision propped up by masculinist-militarized technovisions of control that see nonconforming bodies, land, and collectivities through colonial fears of Indian resurgence.

Captivating Visions: Settler Technoevolutionary Belonging

Visuality and settler colonial occupation are tethered through the scientific, technological, and legal knowledges that produce the power to "see" and "unsee" land and bodies. For example, the English Common Law doctrine of discovery facilitated settler theft of *terra nullius*, or empty land through the proclamation of footprints onto seemingly unclaimed territory, turning the settler colonial gaze into a *scientific weapon*. The very project of seeing and observing the natural world, as argued by Jodi Byrd, as either overly full of Natives or empty of Natives, entitled settlers to ownership of Native land.[23] Once a territory could be proven as "discovered," witnessed as a wild space in need of Indian removal or void of human (Native) presence, it could legally become another's property.

This doctrine borrows from the Roman legal doctrine of *res nullius* (nobody's thing), in which unowned objects without legal rights—wild animals, earth, water, and even slaves—could become property once captured. Subjects within law, such as citizens and propertied land, were excluded. Curiously, when an animal (or slave or Indian) broke free of captivity, they could then recover their natural albeit precarious liberty.[24] Bees, for example, could not become property until hived, or domesticated, as was true for slaves—further entrenching the boundaries between civilized domesticity, law, and rights, on one hand, and wild or unruly spaces and human-animal natures, on the other. The civilizational impulse to own property was premised on containment, the contained zones of the plantation, the farm, animal husbandry, and the reservation—and later the border.

These legal doctrines reveal the socioeconomic and political motivations compelling the rise of Western science as a method of seeing that segregates the viewing subject from the objects it sees and disaggregates the "objects" of the natural world from the human. And the monitoring and containing of one's property—from slaves and Indians to animals and land—ensured one's rights as a citizen. As argued by Ethnic Studies scholar la paperson, "The 'humanity' of the settler is constructed upon his agency over land and nature."[25] Put another way, control over nature, and thus land, was the passport to humanity. Through alienated seeing, settlers participated in the mastery of state vision by "harnessing of nature and its 'natural' people."[26] By participating in state

technologies of seeing, embedded within evolutionary stories of racial development, white settlers ran from the wild natures of racialized others as a past that
started to feel like, literally, another country. The race was on to tame, control,
and take as much land as they could.

H. L. Morgan, Technology, and the Evolution of Intelligence

The US purchased the southern territory of Arizona from Mexico in 1853 as
part of the concession following the Mexican-American War. The military
covered up the mass genocide of Native Americans through scientific studies
that proved Indians were at the origin of anthropological time, marking the
rise of modern human historical evolution. Nineteenth-century theories of social and cultural evolution moved away from hereditary theories of genetics to
posit instead a materialist portrait of time propelled by humans' agential role
as laborers (who turn nature into objects). The most prominent ethnologist
of such a hierarchical evolutionary social order, Lewis Henry Morgan, popularized evolutionary theory in his 1877 book, *Ancient Society*. Morgan aimed
not only to map and archive the history of man and his contributions across
time and space but also to prove that human intelligence could be measured
by the impact their innovative objects have on human society. The earliest
stage of time began with Indians, who created the simplest tools (savagery);
humans then evolved to make more effective inventions (barbarism) and then
finally reached the highest state (civilization), evident in the sophisticated technological innovations of white men. Like most scientific disciplines of his
time, Morgan's comparative approach to racial difference drew from his and
others' studies of Native tribes such as the ancient Aztecs and from the Iroquois Indians whom he studied in person during the late 1800s.

Morgan's obsessive collection of physical traces of human intelligence offers
a window into the colonial psyche driving white men's desire to monumental-
ize their position at the highest point of the world's social order. This early example proves a larger point in *Unsettled Borders*: that border technologies, such
as automated sensors or Donald Trump's wall—however ineffective in actually
preventing the flow of migrants—constitute an archival trace, a monument, or
a footprint of white civilizational superiority and belonging. For Morgan, the
more significant the technology to human flourishing (leading to more widespread architectural designs, patriarchal family structures, private property,
and democratic governance), the higher the group would land on this scale.

Morgan was undoubtedly influenced by Charles Darwin, who established
a similar hierarchy of animals including bees, whose position in his taxonomy

depended on the complexity of their honeycomb construction (chapter 4).[27] Morgan states, "With the production of inventions and discoveries, and with the growth of institutions, the human mind necessarily grew and expanded; and we are led to recognize a gradual enlargement of the brain itself, particularly of the cerebral portion."[28] Folded within this story was a comparative sketch of behavioral evolution that converted intelligence from an innate trait of biology into man's agential role in the conversion of nature into material objects. By measuring intelligence through objects, acumen could be appropriated as a raw material for the next stage of humans to incorporate into subsequent designs. One technological feat was replaced by the next, just as one tribe of people, or species of bees, was replaced by a more advanced one. Archaeological objects defined the present-absence of a people whose tenure on land (mysteriously) expired when they were replaced by a more technologically advanced group. Within this evolutionary model, Native Americans lacked humanity not only based on their designation as savage (lack of intelligence owing to proximity to nature) but also based on their incapacity for creative labor (or ability to turn nature/land into objects). As Morgan argued, those with more advanced labor skills were better equipped to relate to land as property, since they converted nature into objects, which became property. Hence, we see the ways nature had to be separated from objects in the colonized worldview as a precondition for labor/agency, intelligence, and property and thus for becoming fully human.

By wresting intelligence from the immutable confines of biology as well as from the divine design of higher religious forces, Morgan's evolutionary materialization of time and human (creative) labor supports, even as it departs from his contemporary Karl Marx's materialist theory of labor. Early man's preindustrial production of objects might have been simple, leading to modest improvements in society, but this labor, argues Morgan, does not dull the mind. On the contrary, creating material objects out of nature fostered human intelligence and uplifted humans from other animals. It is no surprise that Marx was a great fan of Morgan's book.[29] Even as Native people's savage intelligence was the vital foundation for modern man's eventual rise, Marx's revolutionary sociality found good company with Morgan's depiction of preindustrial labor as an activity associated with creativity, humanity, and intelligence. Neither criticized the racial consequences that propped up this evolutionary drive of historical materialism (and neither could imagine or grasp Indigenous intelligence with regard to land as the creative source for expanding human cognition and hence evolution). Marx fetishized precapitalist labor relations as an alternative utopia to the dehumanizing effects of capitalist industrial production that alienated humans from their creative potential and thus intelligent evolution. Even for Marx this tale of human-technological development also presumed that Indians were stuck in the past and—similar to animals, plants, and the natural world—served man as a raw material for a revolutionary social order that privileged a return to creative labor.

Current innovations in autonomous intelligent robotics extend the evolutionary myth that technological objects make humans smarter. For instance, a recent Alexa ad touts, "Smart Speaker. Smarter Human."[30] Alexa is another version of the Indian scout given that "she" performs everyday reconnaissance missions (yes, sir!) by reporting back information for humans on command. "She" is the object for whom "humans" become more intelligent and thus more human (she is, after all, *merely* a machine). Human intelligence, quality of life (rapid information retrieval), and our very humanity improve alongside our techno-objects. And each data trace is archived and recorded, easily turning our virtual pathways into surveillance capital-as-property.[31]

Morgan felt it urgent to document the traces of each group's contributions before they were replaced, forgotten, and lost to history forever. In fact, media scholar Brian Hochman interprets this drive to document and categorize human behaviors and culture as an innovation driven by the racist beliefs of the nineteenth century, when "writers and anthropologists believed that historical forces had pushed the world's primitive cultures to the brink of extinction."[32] While other scholars have written extensively on the popular perception of the "vanishing Indian" driving anthropology, ethnology, and so forth, Hochman argues that the imperative to archive Indianness—Indians' voices, languages, and culture—inextricably linked extinction and obsolescence to the innovative drive of ethnographic documentation.[33] In other words, Hochman's coupling of racial and technological obsolescence is crucial to unpacking not only how technologies produce racial meaning but also, perhaps less noted, how racial alterity itself motivates innovative techniques, such as phonography, to capture its inevitable disappearance.[34]

By naturalizing the inevitable extinction of racial otherness, automated technologies escape accusations of labor exploitation, military violence, environmental destruction, and settler dispossession. For instance, Tung-Hui Hu exposes the colonial infrastructure that props up the supposed ephemeral placelessness of the internet's digital cloud, stored in thousands of environmentally disastrous data centers and routed across fiber optic lines that trail nineteenth-century telegraph lines, built close to old railroad tracks.[35] The construction of telegraph lines brings us to another example of how technological development participated in the settler theft of Native American relation to land. Literary scholar Kay Yandell argues that these telegraph lines justified the settlement of the western frontier by replacing Native Americans' ability to communicate across space ("moccasin telegraph") with an electrical wire (developed by Samuel Morse) that could transmit the "nation's origin myths and thus sacralize settlers' connections to American lands."[36] This virtual form of communication fostered a disembodied relation to land that attempted to replace Natives' storied and spiritual connection to land.

Also disappeared from technological infrastructure and products are workers: from Chinese railroad workers during the nineteenth century, to contemporary and future Mexican laborers replaced by automated arms that will pick our produce, to mostly female Navajo workers at a microchip facility in New Mexico hired from 1965 to 1975.[37] Not only are laborers racialized and gendered as devalued and hidden labor for high-tech products and functions, but, as argued by Neda Atanasoski and Kalindi Vora, automated machines like robotic laborers also take on a surrogate relation to the freedom and humanity of the privileged class, who can disaggregate human activity from dehumanized and racialized labor—the mundane, dirty, and meaningless labor now performed by robots.[38] Projecting a future in which technological objects replace exploitative and slave-like labor would mean imagining a time, Atanasoski and Vora argue, without gendered and racialized laborers—human servants, slaves, and caretakers—who historically have performed this labor.

To combat the structural imaginary of enslavement undergirding Western manipulation of techno-objects to serve the will of humans, a collective of Native scholars speculate a world in which autonomous intelligent objects are treated not as an alien other but as kin. Against the interpretation of such techno-objects as void of responsibility and relation, and extending Donna Haraway's notion of "making kin," they ask, what if we treated all objects as kin?[39] Through this provocation, they refuse to treat automated objects (as Westerners do) as nonhumans void of interiority and thus as slaves, or machines unworthy of relation.[40]

Surveillance Studies

The field of surveillance and security studies gained prominence and momentum after the attacks on September 11, 2001, by foreign terrorists who penetrated US borders. As part of the effort to secure the nation, attention refocused on the US-Mexico border. In *Innovation Nation: How America Is Losing Its Innovation Edge, Why It Matters, and What We Can Do to Get It Back*, security advisor John Kao argues that 9/11 broadcast a known but unspoken fact, that the United States' inability to innovate at the level required to control its borders was a sign of the country's slow decline as a global power since the Cold War.[41] Kao worked with US national security agencies and departments as an adviser during the 1990s to discuss how to best protect national interests now and into the future through high-tech innovations. We must, he argues, keep an *eye* on future threats *and* opportunities, by identifying the unknowns well before they emerge.[42]

Scholars have disputed Kao's position, arguing that 9/11 was not *the* exceptional event launching us into an elevated state of national security. They claim, rather, that 9/11 brought back Cold War–era racialized suspicion and a windfall of funding for, and interest in, militarized technologies from World War II, such as the drone.[43] Immigration scholars have similarly linked the militarized border to the Cold War, when fears of foreigners and immigrants justified more ubiquitous security against an elusive enemy, while locating a new site (the border) for the use of war technologies.[44] While clearly both the Cold War and 9/11 are important moments in amplifying border security, they occlude our ability to recognize the longer history of settler colonial ambitions driving border surveillance against a universalizable Indian threat, as well as the logic of occupation and dispossession of Indigenous land. Against these erasures, Native American scholars understand 9/11 as a national threat haunted by Indigeneity, evident in the military coding of Osama bin Laden as Geronimo.[45]

As the field of surveillance gained traction post-9/11, the majority of scholars in the 2000s popularized an analysis of state surveillance as emerging with the rise of modernity, most readily evident in Michel Foucault's theorization of the panopticon. For Foucault, this "watchful gaze of the state" became a key technology to follow broader sociopolitical transformations in state control, from spectacular displays of torture and violence to modern bureaucratic discipline, from the guillotine to the watchtower, and from a sovereign ruler empowered by the divine right to kill to the state's bureaucratic violence of containment. This story of the state's techno-evolutionary progression—from bloody brutality to bureaucratic discipline—obscures the ongoing colonial violence hidden within the surveillant eye that captures and transmits data. For Paul Virilio, the watchtower was more than a disciplinary tool; it was a visual weapon of war: "from the *original* watch-tower through the anchored balloon to the reconnaissance aircraft and remote-sensing satellites, one and the same function has been indefinitely repeated, the eye's function being the function of a weapon" (my italics).[46] Contra Virilio, Caren Kaplan questions the accuracy of aerial views, however weaponized, and destabilizes this totalizing viewpoint by attending to what cannot be captured or known by the eye in the sky.[47]

Few scholars consider the colonial use of *vigías*, or watchtowers, erected by Spanish colonizers to monitor the movements of American Indians in the Southwest and Maya in the Yucatán. These early towers aimed to parse land and bodies into the governable and the ungovernable, or the wild and the civilized, as well as the licit and the illicit, especially by controlling the trade of unauthorized goods, as well as preventing the Maya from escaping into the jungle, out of the reach of colonial tribute and control.[48] The surveillant eye of the *vigía* serves as a colonial technology of bordering, of territorial dispossession and bodily containment, demarcating racial and territorial boundaries while determining the jurisdictional control of Indigenous movement. For the O'odham, border security today extends the violence of governance from the discipline of the watch tower to an atmospheric occupation aimed (unsuccessfully) at Indigenous extinction. To tear apart their reservation through the imposition of a border is to attempt to render extinct their sovereignty and freedom, which are inextricable from movement, from following the footprints of their ancestors that preserve the lively connections with the land.

As many have noted, Foucault's limited analysis of racially targeted state violence, confined to the horrors of the Holocaust, led him to see racialization as an exceptional problem found elsewhere and as a historical event in the past, sending many scholars down a similar path of forgetting the centrality and ongoingness of the racial containment of slaves onto plantations and of Native peoples onto reservations.[49] Or as Macarena Gómez-Barris asserts, theories of visuality that begin with modernity forget the context of coloniality and thus "render invisible the enclosure, the plantation, the ship, and the reservation."[50] And, I would add, the border. What all of these spatial enclosures have in common, despite their distinct temporalities and territorial configurations, is a biopolitics that extends beyond simply disciplining bodies through containment but of investing in the ongoing slow death, or extinction, of Black, Latinx, Asian, and Indigenous life. If we see the rise of surveillance alongside land-based struggles like the Indian Wars from the 1700s to the 1880s, the quest for land goes hand-in-hand with settler surveillance as a military tool for expanding state sovereignty and control.[51] This happens in part when we focus, as Andrea Smith argues, on what surveillance sees, what it makes hypervisible versus what "delegitimizes the state itself."[52] This critical framework of what is visible or hidden is helpful in understanding the book's focus on how Indigeneity drives the technological innovations of the settler state and attempts to assert its sovereign power.

Feminist and critical race perspectives on surveillance have pushed the field from Eurocentric perspectives on the rise of modernity to critical accounts of the racial underpinnings of state seeing.[53] Rather than focusing on the bodies caught in the surveillant eye, it is imperative to situate settler seeing within the historical and scientific context in which surveillance emerges, including the settler military imaginaries that continue to hide how this gaze targets Indigenous bodies and land, transplanted globally to mean any foe that threatens the colonizers' land-based power. To interrogate surveillance as a settler colonial technology is to attend to the ways Western spatial imaginaries desacralize relational bonds between land and bodies, regarding them instead as unruly entities to be conquered. The border's spectacular resonance as an uninhabitable or wild frontier—either empty or overrun—continues to justify the need for settler presence and a ubiquitous security apparatus.

The fields of feminist science and technology studies have also emerged to combat the growing militarization of knowledge resulting from alliances between the Department of Homeland Security, the Defense Advanced Research Projects Agency (DARPA), and universities. As public state funding diminishes for higher education, these agencies offer students, researchers, and faculty ample grants, funded centers, and endowed positions for becoming the research and development arm of border security. The University of Arizona, just an hour from the border, is one of the leaders in this field. Such alliances demand we interrogate how academic theory supports the logic of these innovations in science and technology. By attending to the technoscientific worldviews driving surveillance, we move away from a focus on stable subordinated identities to the very processes of subjection, or from seeing racial bodies to unpacking what Ruha Benjamin defines as the discriminatory design of the state and its technoscientific methods that produce racialized ways of seeing and knowing.[54] If we return to the historical context of military surveillance, these technologies were designed not only within a racial imaginary that sees the Indian as a threat. For within the evolutionary context of automated intelligence, technological advancements in border security led to infrastructural occupation on Native land that threatens to render Native peoples obsolete, or extinct.

State surveillance continues scientific and neoliberal rationalities that encourage us to believe that governance operates best when it can see problems with a distant gaze and when it develops technologies that provide more humane, efficient, and accurate vistas. The surveillant gaze reinforces the belief that the phenomena we can see and know are good and that what we cannot see is false/deceptive, dangerous, suspicious, and threatening, resulting in a neocolonial tracking system that justifies the need to light up, or see in the dark, to see what cannot be knowable or fully controlled. In her book *Dark Matters: On the Surveillance of Blackness*, Simone Browne traces the racial underbelly informing Foucault's theory of biopolitical governance through the surveillant panopticon to Jeremy Bentham, who came up with this term while on a slave ship. By centering slavery in the heart of surveillance, Browne argues that surveillance produces definitions of Blackness (what cannot be seen or known and thus is in need of control), while at the same time anti-Black racism prefigures the need for, and fabrication of, these technologies. When one traces this earlier history of surveillance, the panopticon fails as a register of modernity and its association with bureaucratic efficiency, objectivity, and unbiased development.

Through a present-future secured through ubiquitous surveillance, these technologies target racialized bodies and extend the state's intrusive eye into more areas, reaching beyond its sovereign control.[55] The act of seeing like a state thus weaponizes illicit bodies, land, and knowledges, distinguishing the safe from the dangerous, the white from the racialized, the normal from the abnormal. Attempts to visually produce evidence of the racialized body as a criminal threat or vector of disease at the border can be seen in the rise of technologies used to identify certain migrants. For example, in the late nineteenth century, Chinese migrants were the first to have photographs taken of them at the border (at a time when photographs were used as proof of criminality), and Mexican migrants in the early twentieth century were quarantined, sprayed with DDT (thought to be a delousing agent), and branded with an iron to mark them as "authorized."[56] What interests me is how this visual technology automates these racial stereotypes into data visible on the body and land to justify the need to visually map every dangerous movement and unstable region bordered by surveillance.

Refusing Technoscientific Seeing through Apache, O'odham, and Maya Sacredsciences

National border control fantasies of multiscalar surveillance, of seeing from every dimension possible (with the goal of controlling more space), stand in stark contrast to Native sacredsciences that have long observed the shape and intelligence of humanity in motion with the multidimensional spirit of life. By interweaving science and the sacred, I aim to highlight Indigenous perception as a technology honed across generations that extends human becoming and intelligence with the more-than-human world. Sacredscience consists of practices and knowledges that re-enliven relations with ancestors and that activate a calling to nourish the intricate web of entanglements that sustain all living beings on earth. Compared with the objective approach in Western science that alienates one phenomenon from another in order to control the natural world through ubiquitous vision and, thus, knowledge, sacredscience refers to a knowledge system that fosters one's responsibility to respect the relational web and life force that holds us all together, glimmering in a dimension that eludes Western systems of knowledge.[57]

Thus, against the military theft of Nativision tied to the land, *Unsettled Borders* follows the more-than-human knowledges, fugitive movements, and acts of freedom that exceed technologies designed to control, incarcerate, and limit Indigenous land-based flourishing. These opposing forces have each gained momentum, and the tension between them has reached a feverish pitch. As Native voices rise up with greater force, advances in science and technology promise to unlock the mysteries of life and gain greater control over it. I argue that describing Native American cosmologies with place as a sacredscience is critical at this particular moment. As noted by many Indigenous scholars, Native knowledge is often considered as the antithesis of science: as myth, primitive belief, or as raw material to be extracted and developed through Western science, patented as its own, and commodified.[58] Growing movements of Native activists are refusing settler development on their sacred land, from Standing Rock to Mauna Kea, and at the US-Mexico border by the Tohono O'odham, who protect their sacred burials and springs from destruction by the construction of President Trump's border wall (chapter 2 and conclusion). In the process of refusing the construction of an observatory on Mauna Kea, a mountain the Kānaka Maoli consider sacred, Native Hawaiians are accused of being antiscience, an epithet with a punch given the centuries of denouncing many Native peoples for their "primitive beliefs" unsupported by scientific methods such as documentable evidence.[59] A 2014 *New York Times* article labeled Indigenous creationism as opposing science, and opposition to the Mauna Kea telescope as a "turn back to the dark ages."[60] To understand the contestations over Mauna Kea, Hi'ilei Julia Hobart traces the "superimposition of Western spatial imaginaries—particularly emptiness—upon Indigenous geographies [that] has been used to justify a number of development projects."[61] Not only is the land emptied of Native Hawaiian presence but, Hobart continues, the land is also deanimated from a Kānaka Maoli perspective that sees intention, ancestor, and spirit when a deluge of snow from the mountain halts all construction and settler passage up to the summit.[62]

In the case of the San Carlos Apache, who stand against the University of Arizona's construction of two telescopes on their sacred mountain, Dzil Nchaa Si'An ("big seated mountain"/Mount Graham) in Tucson, Arizona, anthropologist Elizabeth Brandt reported that proponents of the telescopes wanted physical proof of "sacredness," such as extensive ruins, a temple or church, or a burning bush, evidence that falls outside of Apache understandings of the sacred as a way of living with land.[63] Misunderstanding the sacred through Western concepts of religious freedom, defined as the freedom to practice one's beliefs in places of worship (such as a church or mosque), has turned a blind eye to Native peoples' sacred relation to land. As argued by Winona LaDuke, "Some 200 years after the U.S. Constitution guaranteed freedom of religion for most Americans, Congress passed the American Indian Religious Freedom Act in 1978," and while this law protected the legal rights of Native peoples to hold ceremony, LaDuke continues, "It did not protect the places where many of these rituals take place or the relatives and elements central to these ceremonies."[64]

In 1883 the Department of the Interior declared Native American religion illegal, attempting to destroy Native knowledge, power, and relation with land.[65] As scholars such as Keith Basso acknowledge, various Apache tribes find meaning, memory, and knowledge in place, or "place-based thoughts."[66] Nicholas Laluk, a white mountain Apache anthropologist, finds that the most important word in the Western Apache language is *Ni*, which means both "land" and "mind."[67] Knowledge and thought are not the sole domain of humans but are inspired by, and inseparable from, the intelligent design of all scales of life. And not only do cognitive and social meanings reside in relation to place, but places are spirited, alive, and part of the meaning-making process. Thus, to track footprints is a key methodology in this book, not only for following the past but also for seeing with Apache and many other Native-inspired place-based knowledges. Anthropologists such as Laluk, as well as Lesley Green and David Green, find that archaeology imposes a Western framework on Indigenous expressions, or, better yet, practices of history, time, and place.[68] In contrast to looking to archaeological objects to determine "belief systems" or historical "presence" (as if Native peoples and their ancestors did not still reside in those areas), "footprints" or "tracks" offer an alternative worldview to think about what it means to follow the past as a relative that guides one's actions. Western understandings of "history" reserve it as a temporal placeholder that alienates the past from the present-future, while objectifying time as independent of human and nonhuman intervention. Basso describes the ways Chiricahua and other Apache experience places, rather than objects or time, as holding memories that are awakened, and storied, by ancestors when one follows their footprints or tracks on the land. For example, the time of flooding in the Huachuca mountain desert region is told by the naming of a place or by stories that describe the larger events accompanying the mineral residue of water imprinted on the rocks. This event holds importance owing to its relevance in the present. Learning the footprint of water on a rock or the motion of the stars and moon across the sky provides lessons on how those movements relate to the annual rising and falling of the waters, or the change of seasons, the timing of when to plant, and the migration of humans and animals.[69] What appears to be a "localized" knowledge allows one to understand the entanglement of cosmic or local events. To see the great migration of settlers from the east—the tracks of the white people and their tools, animals, bodily movements, eating habits, and technologies—is to understand how these movements will alter the world around them. To learn to see how land wilts or dies off, how it shows signs of lifelessness when overworked by extractive industries, monocultural factory farming, or other kinds of misuse is to foresee a time to come that teeters toward destruction if we don't take responsibility to repair these wrongs. By tracking footprints, one can come to understand, and take part in, the artful dance of life and death of a world in constant motion.

Terms usually thought of as antithetical—*sacred, technology,* and *science*—are juxtaposed in the book to resignify, reimagine, and refuse Western hierarchical racial orders guiding science and technological development. Sacredscience is a term in conversation with a range of Xicana and Indigenous scholars who highlight land-based ontoepistemologies, what Michi Saagiig Nishnaabeg scholar, writer, and activist Leanne Betasamosake Simpson calls Nishnaabeg intelligence, or land-based pedagogy.[70] Through sensual stories of maple harvesting told with childlike awe and love, and drawn from deep observation with squirrels and trees, Simpson situates harvesting as an intelligence "woven within kinetics, spiritual presence and emotion, it is contextual and relational."[71] Through this traditional story Simpson shows how technological "discoveries" come out of love and respect for the land, thereby strengthening the communal bonds between people who together take part in the creative act of feeding the mind, belly, and social relations in ways that strengthen the bonds connecting people with land. Simpson knows that "settlers easily appropriate and reproduce the content of the story . . . when they make commercial maple syrup in the context of capitalism, but they completely miss the wisdom that underlies the entire process because they deterritorialize the mechanics of maple syrup production from the place, and from freedom with Aki, or land."[72]

These stories remind us that Nishnaabeg "discoveries" strengthen their freedom to uphold collective responsibilities.

In a similar vein, the Tohono O'odham refuse their extinction by surveillance, extraction, and dispossession of their land, by following in the footsteps of ancestral ways, including communing with their creator/mountain through songs of prayer and harvesting sacred plants such as the saguaro cactus whom they consider to be an ancestor that teaches many lessons on how to live, survive, and thrive in the desert (chapter 2). The Yucatec Maya turn back to ancestral practices of beekeeping, to remember knowledges that not only counter the militarized extraction of land, outmigration, and dispossession but strengthen relationality with each other and the land (chapter 4). As a sacredscience, these ancestral knowledges entail seeing with ancestral eyes that preserve knowledges across timespaces that aid the present. Past knowledges reside in the plants, animals, and people that engage in these practices to ensure futurity for many to come. Indigenous knowledges are not simply accumulated traditional ecological knowledge (TEK), bound and archived, that can be picked up by scientists and environmentalists and transferred to any location, as noted by Potawatomi philosopher Kyle Powys Whyte.[73] Instead, he argues, they are situated knowledges embedded in the worldviews, creation stories, and cultural practices of a community of people. Held within these sacredsciences are memories that kinetically regenerate people's freedom to move with the laws of the many creatures inhabiting the entangled, yet distinct, worlds of their land. Anishinaabe speculative fiction scholar Grace Dillon skeptically uses the term *Indigenous scientific literacies* as critical to Indigenous futurity: "In contrast to the accelerating effect of techno-driven western scientific method, Indigenous scientific literacies represent practices used by Indigenous peoples over thousands of years to reenergize the natural environment while improving the interconnected relationships among all persons (animal, human, spirit, and even machine)."[74] Ancestral sacredsciences are remembered as footprints that inform the present-future, while settler science looks to a future in which the past is overcome, or improved upon.

The scholarly work of Vine Deloria, one of the leaders of the field of Native American studies, addresses the strength of Native spiritual knowledges that were delegitimized as worldviews and methods because they threatened the legitimacy of Western science. Deloria convincingly contends that Native knowledges surpass Western knowledge systems, especially since they do not rely on one single thinker or scientist and instead gather wisdom from a diversity of intelligent perspectives across the human and nonhuman and across time. Deloria says, "Indians consider their own individual experiences, the accumulated wisdom of the community that has been gathered by previous generations, their dreams, visions, and prophesies, and any information received from birds, animals, and plants as data that must be arranged, evaluated, and understood as a unified body of knowledge."[75] Robin Kimmerer, a Native ethnobotanist and citizen of the Potawatomi Nation, argues that Indigenous TEK should be seen as parallel to science. The two major differences are that TEKs are qualitative, based on observations by a community over a long period of time and that, unlike in Western science, with TEK, the observers tend to be those who use the resources themselves and thus consider what is observed from the position of subject rather than object. Kimmerer reminds those who protect the land without regard for the inhabitants that environmental problem solving cannot be disconnected from human values and worldviews.[76] Gregory Cajete, a Tewa scholar, goes even further to argue that Native science rejects the objective distance of Western science and instead acknowledges the material and spiritual ties of transformation when he says, "in learning from and eating each other, we are transformed into each other."[77] In a similar vein, the physicist F. David Peat sees connections between quantum physics and Native philosophy, which both recognize an experiential and ontological shift: "within the Indigenous world the act of coming to know something involves a personal transformation. The knower and the known are indissolubly linked and changed in a fundamental way."[78] The book extends these ideas to consider how becoming animal (both biologically and imaginatively) is a technology that extends human becoming, knowledge, and perception of the world through each animal's particular aptitudes.

These sacredsciences come from everyday necessity and collective visions, pushing biological knowledge into an entire sacred cosmology that accounts for the lively presence and knowledge of wind patterns, animal habits and behaviors, the edible parts of animals and plants, plant growth, star patterns, and what all life forces need to survive, propagate themselves, and thrive.

Alongside Native scholars and scientists and feminist science and technology studies scholars, the book draws from decolonial Indigenous, Xicana, and queer methods to address realms that fall outside rational technoscientific perspectives, such as embodied knowledges that come through dreams, visions, poetry, conversations with elders, experiences of birthing and dying, and ceremony. To follow footprints entails seeing the not-quite-visible through other animal-plant-star movements and stories, songs, and visions with ancestors. I am especially moved by Gloria Anzaldúa's erotic cosmontology, or sensual becoming through deep engagement with the land, or the more-than-human world, that builds consciousness, or knowing and becoming beyond the limits of the human. I see in Anzaldúa's writing a deep engagement with the spirited forces all around her, a methodology of perception in which she not only intimately learns to see with the more-than-human world but becomes other with them. This third-space methodology in Chicanx/Latinx decolonial scholarship builds alliances with Indigenous cosmologies that refuse to separate human from nonhuman perceptual ontologies, a separation foundational to Western borders between the self and other. In a similar vein, Macarena Gómez-Barris builds on a decolonial queer episteme that privileges an Andean phenomenology that respects sensual, intimate, and embodied relations with land, to break with the dispossessive logics of extractivist perceptions driving spiritual tourism in Peru.[79]

Indigenous and feminist science and technology studies scholars such as Kim Tallbear and Angela Willey theorize sexuality as a potential technology for engaging science otherwise, breaking the heteronormative hold on scientific ways of knowing by identifying the healing potential of erotic intimacies that transcend sexual reproduction and the human.[80] Tallbear evocatively asks us to consider how polyamorous relationalities might expand our kin networks and enact collective healing. In a related vein, queer literary scholar Mark Rifkin turns to queer Indigenous writers such as Qwo-Li Driskill to articulate an erotics of sovereignty that decolonizes empirical and state definitions of "the real." Instead, Rifkin argues that embodied knowledges with land bring back sensually perceived relatedness that reignites felt moments of freedom that disappear the bounds between human and nonhuman, self and other.[81]

There are other decolonial scholars that are important to acknowledge as having similar goals of expanding what Arturo Escobar calls the One World approach, universal knowledges that replace the detached perspective of science and the academy with multiple perspectives, or a turn to the relational knowledges that lead to many worlds, or what he calls the Pluriverse.[82] There is a group of decolonial, Indigenous, and postcolonial science and technology studies scholars whose methods dovetail with the perceptual apparatus from Indigenous and subaltern perspectives from below, including what Macarena Gómez-Barris calls "submerged perspectives," and Zoe Todd calls seeing with "fish pluralities."[83] There is also an emergent group of feminist science studies scholars who attempt to incorporate postcolonial approaches to see the world from perspectives that refuse to hierarchically categorize life with the human at the apex. For example, Deboleena Roy asks what kinds of becomings might be possible when we see beyond the human, from and with grass.[84]

These scholars enact theories from places and from intimate knowledge of place that emerged through respecting the intraconnected patterns of life that affect us all. Indigenous peoples across the Americas refused to submit to the Spaniards' sovereign God that transcended the human earthly realm, or, later, the sovereignty of colonists and their allegiance to a king or nation ruled by humans who thought themselves superior to nature. An alternative understanding of sovereignty based on respect for the flow of life's infinite forms—from the sun to rivers to plants—challenges settlers' belief that nature was a wild sphere to be domesticated through borders on a map demarcating private property, which in their mind and practices segregated wild from cultivated land. As forcefully stated by Val Lopez of the Amah Mutsun tribe, it's not that Native peoples did not alter their landscapes, but they did so in ways that fostered the growth of plants, creating water systems that were used by Indigenous peoples without cutting off the flow of the fish, and so forth.[85] The stakes of the project entail both understanding the constant yet changing forms of settler colonialism that continue to push the bounds of state sovereignty and also heeding the call of Native communities who risk losing their land and their particular way of knowing and being in the world. When people's land is stolen, occupied by the military, destroyed by extraction and toxic dumping, what webs of relational becoming disappear or go extinct?

Chapter Overview

In each of the chapters, I track the Native footprints driving settler surveillance as an automated structure of seeing that is extractive, removing and containing threats to settler governance, capitalism, and belonging. This colonial structure of seeing bodies beyond state detection as dangerous intruders reinforces military occupation, carceral containment, and elimination and dispossession of Native land and bodies as well as migrant border crossers. At the same time, each chapter interrupts these violent ways of seeing through demands for Indigenous autonomy maintained by outright protest, as well as by less understood collective relational practices with land. By elucidating the tensions between automation and autonomy, I hope to unravel the binaries foundational to Western knowledge and subjectivity that continue to drive the violent settler colonial and extractive desires of the state, while pointing us to other ontoepistemological possibilities that support and proliferate life on earth for all.

Chapter 1, "'The Eyes of the Army': Indian Scouts and the Rise of Military Innovation during the Apache Wars," situates military surveillance as an innovation developed within the laboratory of the Apache wars. As a tool of settler violence, Native "eyes" or visionary skills had the power to aid or disorient settler military control of the Southwest border region. While Indian scouts were dubbed "the eyes of the army," Apache visionary practices with land were tied to centuries of adaptations to all the forces of land I call a *sacredscience*. Their animated vision and communication across space constituted a powerful tie to land that was dangerous to military-backed settler belonging. Early military innovations, such as the heliograph and binoculars, were created to extend settler seeing in order to track down, contain, and replace Apache fugitives in frontier regions considered remote, wild, untamable, and hostile to the civilizational might of settler presence.

In chapter 2, "Occupation on Sacred Land: Colliding Sovereignties on the Tohono O'odham Reservation," I move from the Apache Wars to the current border war waged on the Tohono O'odham reservation. While the O'odham have no word for "border," the United States declared the region between Arizona and Mexico a security "void" to justify their illegal invasion of the sovereign O'odham reservation. This chapter examines the colonial history of Western visual mapping of O'odham desert land as inhospitable or empty to contextualize the current military occupation of the reservation by infrastructures of security. Given that little scholarly attention has addressed the Native peoples living on the border today, this chapter assesses how the same immigration policies and infrastructures that funnel Latinx migrants and smugglers onto O'odham sovereign land—many of whom die crossing the desert— target O'odham tribal members. Many O'odham tribal members and activists see this militarized intrusion as another attempt to occupy their land and render them extinct, or to dismantle the sacred worldviews and practices that make them a distinct people. Similar to other sovereign nations along the US-Mexico border, they demand autonomy and an end to all incursions by border-security technologies, from the wall to surveillance towers, that close in on them from the ground to the sky.

Chapter 3, "Automated Border Control: Criminalizing the 'Hidden Intent' of Migrant/Native Embodiment," traces the long legacy of collaboration among the state, the military, and universities that has led to the branding of Arizona as Optics Valley, a laboratory for visual technologies that include surveillance, drones, sensors, and observatories. In this chapter I trace the research and development that created the sensors used to scan border travelers in the automated border kiosk AVATAR, funded by the Department of Homeland Security and developed at the University of Arizona. Automated border surveillance today promises to visualize and detect threats across all scales— from the ground to the sky, including inside the body. The AVATAR kiosk invades the body with fifty sensors that track the hidden signs of physiological deception with the goal of identifying whether people are crossing the border for good versus malicious reasons. As the body is turned into matter, biological life resurfaces as a more scientifically verifiable truth than human verbal testimony, moving border security into the colonial recesses of the body's unconscious movements. This chapter unpacks the connections between the nineteenth-century visual capture of "marauding Indians" and the detection of a primitive Indian racial unconscious that continues to inspire Arizona's border-security-industrial complex.

Chapter 4, "From the Eyes of the Bees: Biorobotic Border Security and the Resurgence of Bee Collectives in the Yucatán," turns to another phase of automated security that draws from scientific studies of "swarm intelligence." I trace the turn in autonomous intelligence from human to animal intelligence through nineteenth-century debates about bee intelligence in the natural sciences, such as the views of Charles Darwin, to consider how the rise of biological sciences inaugurated the move away from religious, spiritual, and any other unseen life forces, a materialist debate I track through scientific observations with bees. The last section of the chapter disproves the evolutionary disappearance of the Maya/bee by focusing on the resurgence of beekeeping by Maya in the Yucatán, Mexico. Through ancestral sacredsciences tied to beekeeping, women engage in land-based protests against the containment and disappearance of their land by logging companies and tourism economies. Through the campaign "I am the Maya bee," they refuse to separate collective demands to protect the bees, their land, and women's bodies from the violence of extractive industries, environmental devastation, the murder of women eco-activists, and the loss of bees. In addition to demanding autonomy, Maya build sustainable approaches to living that they hope will bring Maya refugees back home.

And last, the conclusion, "Wild versus Sacred: The Ongoing Border War against Indigenous Peoples," returns to Organ Pipe Cactus National Monument on the Arizona desert border, a public wildlife preserve managed by the US Department of the Interior, on land considered sacred ancestral land of the Tohono O'odham nation and the Hia-Ced O'odham. This chapter forges connections between conservation and border control, discussing how surveillance-as-conservation serves as another settler tool to protect the land from human threats, such as O'odham and immigrants, while opening it up to settler tourists. On this land Trump began the construction of his thirty-foot-high metal-and-concrete border wall. I return to the saguaro desert, where the footprints of O'odham, Apache, and Maya collide. O'odham fight against the construction of a border wall—a process which has led to the destruction of some of their sacred burial sites and the extraction of sacred water from Quitobaquito

Springs—to protect not only their history but the sacred footprints of the living and dead: Hohokam and Apache ancestral burial sites, Maya and Latinx refugees and migrants, and a range of ancestral animals, plants, and resources at risk of dying out.

For this book I traveled across the Southwest and into Mexico—from Fort Huachuca to the Tohono O'odham reservation and the Maya territory of the Yucatán—to remap the border from settler eyes to the footprints of the original and ongoing inhabitants of this region. My own family's footprints crossed Indigenous land spanning Northern Mexico and Texas, the state where they settled. This book aspires to critically assess how a focus on the militarized border as it targets immigrants inadvertently erases the perspective of Native border dwellers. I hope the book can thus deepen an alliance and activist claims to justice across these unsettled borders. The land and its original people have spoken to me, and I've tried to listen with my heart's eye. Like Anzaldúa, I sense a shift in form, a new face and time emerging. A shedding of settler colonial skin. A rebirth, an awakening. And as many Xicana and Indigenous writers, poets, and thinkers have prophesized, *the land is and will be Indian land.*

ONE

"The Eyes of the Army" *Indian Scouts and the Rise of Military Innovation during the Apache Wars*

The [Huachuca] post's story . . . is one of a savage contest of arms between dedicated and able frontier army soldiers and implacable Indian braves, a confrontation which culminated in the inevitable reduction of the primitive by the technologically advanced.—Cornelius C. Smith Jr., *Fort Huachuca: The History of a Frontier Post*

The land is powerful and sacred, it holds memory. There is power in those mountains, a power you can't explain. . . . If you return to the Guadalupe Mountains, you will hear the crown dancers and singing in the caves, even though there's nobody there. Ancestor's presence . . . is held in the land. Like a great wind, they left a powerful memory of their passing.—Apache interviewee in Goodwin, *Geronimo and the Apache Resistance*

· · ·

I came across the statue shown in figure 1.1, *Eyes of the Army*, situated between the Fort Huachuca Historical Museum and the US Army Intelligence

Figure 1.1. Dan Bates, *Eyes of the Army* sculpture, Fort Huachuca Museum, 1995. Photo by author.

Museum, a monument that orients visitors on how to interpret the historical connections between the Indian scouts from the 1870s and the high-tech automated surveillance deployed along the border today.[1] With their back protected by the surrounding mountains, the Indian scout points two fingers toward the valley below from a vista high enough to see incoming intruders. The statue performs a disappearing act on settler colonialism and genocide even as it commemorates Apache scouts as heroes in service of the army's defeat of Geronimo, a historical moment that supposedly ended the Apache Wars in 1886. In this chapter I examine the online archival collections and educational displays held within these two museums to consider how and why the US Army showcases Indian scouts as technological extensions of military intelligence and surveillance. Despite Indian scouts' ability to code, or to communicate information across long distances, Native intelligence remains contained within the historical archive as a primitive skill. The statue memorializes the Indian scout's anatomical "eyes," or Nativision, as a natural resource "discovered" by the US Cavalry that expanded Western soldiers' vision into rugged, unfamiliar, and complex terrain.[2] Today the military phrase *eyes of the army* refers to unmanned aircraft systems, or drones, many tested, developed, and displayed at the intelligence museum at Fort Huachuca, with Native American names such as the Apache, Shadow, Blackhawk, Hunter, Mohawk, and Ikhana.[3] Imagined today as abstract code, Native knowledges tied to land were stolen and imprinted into the technological innovations of US settler colonialism and empire. This evolutionary story staged at Fort Huachuca enacts erasure by telling a history of technological extraction that begins with primitive man and ends with automated control of modern techno-gadgets that can track all forms of life from above.

Included in the US Army Intelligence Museum catalogue, "The Intelligence Impulse: A Showcase for U.S. Army Intelligence History," is a section called the "Apache Scout." In this collection the Apache scout is categorized alongside other technologies under "human intelligence" owing to their superior skills in long-range reconnaissance. In fact, Apache tracking skills are described as the ground force that inspired modern techniques from outer space today, such as the "imagery interpreter." They explain, "Imagery Intelligence studies the earth's surface for clues to identify and locate enemy activity. Today that is accomplished mainly by photographic, radar, infrared, or electroptic images, some conveyed from platforms in space. The Apache too scrutinized the ground for signs of enemy activity, but he gathered his images from as close to the earth's surface as you can get. . . . We live in an age of 'Electronic Cavalry,' with new and powerful systems being fielded every year that dramatically increase the commanders field of vision."[4] In fact, a section of the catalog on ground surveillance radar states, "Replacing the Indian Scout of a century ago the three-man Ground Surveillance Radar team gives the maneuver battalion commander a highly mobile, almost all-weather, round-the-clock surveillance of the battlefield."[5] The US Army Intelligence Museum and Fort Huachuca are at the center of a technological revolution in "military intelligence."[6] Similar to the case of abstract or alienated industrialized labor, the catalog converts "Native eyes" from a human skill mastered over time in a local place into an automated universal view from above learned through education.[7] It also converts skills, or intelligence, into technological objects.

The belief that certain Apache and other Native peoples are natural warriors enjoys a long archival presence and is disconnected from the context of settler colonial violence. At Huachuca, this warrior spirit holds sway as a powerful force the US Army attempts to contain and incorporate into an archival past even as Apache skills reemerge in an automated technological future imagined as a war without end. The Huachuca online archive and museum displays celebrate Native warrior intelligence only when it furthers US military interests but easily turn on Indians as a threat when they block expansion, war, capitalism, and thus historical progress.[8] In other words, the Huachuca archives, including the statue in figure 1.1, attempt to incarcerate the threat of Indigenous insurgence through the incorporation of Indigeneity into the US national body politic and narrative as the driving force of militarized innovation. The insurgent force of the Indian warrior emerges in the context of settler incursion, a force the military wishes to harness but one that is misunderstood given that these skills are also an embodied ancestral sacredscience that guides some Indigenous peoples as the best defenders of *their* land.

The repetition of the heroic Indian warrior has deleterious effects for Native Americans today such as contributing to their placement on the dangerous front lines of war or on night-watch duty.[9] In contrast, insurgent practices continue under the banner of a warrior spirit as Native defenders protect their rights and responsibilities to occupied land, or land under threat of pollution or extraction. As Anishinaabe activist and scholar Winona LaDuke discusses, the Ojibwe word for "warrior," Ogichidaa, does not refer to killing but an honorific title and practice for "those who defend the people."[10] The footprints of ancestral warriors are remembered in ways that have material effects, such as driving a mounting force of Native land-based activism. At the same time, the desire to defend one's people and land has sent a staggeringly high number of young Native men and women to join the US military from before the Civil War to today.[11] Of course, many complex factors drive young men and women into war, including long legacies of family separation through boarding schools, a lack of local educational opportunities and meaningful work, poverty and violence, and state campaigns that target Native communities, such as the onslaught of military recruiting on Native reservations. Some see joining the US Army as an opportunity to gain the training necessary to defend their own people and land, or to gain the experience necessary to join the police force or Border Patrol, or to access other perks such as access to education, housing loans, and voting rights, as well as to gain honor and respect back in their home communities.[12]

While the army's selective recuperation of Native Americans as heroes is part of the multicultural liberal story of Indian-army collaboration toward civilizing the frontier, this commemorative veneer erases the violent stripping of Indigenous sacred knowledge or sacredscience. Fort Huachuca resides on the unceded land of various Apache communities at the base of the Huachuca mountain range, or "place of thunder," considered sacred by the Chiricahua Apache and others in the past and present.[13] Revered for its powerful ancestral presence, this mountain range offers gifts of life and of vision.[14] A seemingly unmoving or inert mass to the conquistadors, the US Cavalry, and settlers, the mountain range is to many Native peoples an animate being that could proliferate life or destroy it, a thundering presence that brings life to many of its inhabitants. Many Apache respect the power of this great mountain to hold the spirits of their ancestors and to pull people and animals, as well as wind, clouds, and rain into this dry desert area, contributing to the diverse wildlife and mild climate.

Before I requested a military ID to authorize my entrance into Fort Huachuca, I first drove up to the top of Huachuca to share an offering. I spoke its name and asked for permission to be on this land as well as guidance about the story it wished me to tell. I stood still atop one of its peaks as the wind howled strongly against me, voices from the past breathing stories as they flew past me, filling the sky and land with heartache, laughter, and a heavy feeling of presence. While I didn't see the contemporary gatherings of the San Carlos and White Mountains Apache who return to this land year after year, I heard the muffled sounds of drums and songs echoing off the mountains from far away. Apache return to their land to gather acorns as they have always done and to sing songs of prayer. They walk in the footsteps of their ancestors so the land does not forget them. And the land, wind, and mountains speak back.

With these other presences on the land in mind, I read against the grain of military archives, books, and museum collections written primarily by white cavalry lieutenants and settlers. I also turn to novels, an interview with Geronimo

(1906), and a 1988 documentary narrated by descendants of Geronimo (*Geronimo and the Apache Resistance*) for alternative perspectives on the Apache Wars. For the film, several Chiricahua and Apache descendants were interviewed in the 1980s to share their perspective on the so-called Indian Wars, opening up Apache cosmology and stories that counter the dominant story of this pivotal time.

As told by the Fort Huachuca Historical Museum, the docent, and archival materials, as well as from Apache perspectives, this mountain view bestowed various groups an outlook to see oncoming intruders from all four directions across multiple time periods—Apache; Spanish colonists; Mexican, British, and US soldiers; and other tribes and animals. As I hiked down the mountain, I came across a plaque where Francisco Vásquez de Coronado and his expedition built a watchtower in 1540 to monitor the Indians and to map their voyage across this new land. This mountain peak was also the lookout point that cavalry soldiers partially climbed during the late 1860s–1880s to hunt Apache, especially since the Chiricahua Apache and Sioux moved seamlessly across both sides of the mountain border along the newly won US territory of the Mexican Southwest before and after 1848.

After the Civil War, Congress approved funding for Fort Huachuca, an army post established to capture the last group of Indian rebels and to secure the western border from Indians and Mexicans who fled from their jurisdictional control. They also sought to ensure the safety of settlers traveling and living in the border region. From this vantage point, border control begins not only with the control of Mexicans but with the war against Indigenous peoples on this land. The southern half of Arizona was not included in the 1848 Treaty of Guadalupe Hidalgo; it was surveyed and then purchased in 1853 through the Gadsden Purchase.[15] To wrangle this land from Mexico, the US Army had to first agree that it could prevent Indians from raiding into Mexico. Then they had to prove they would make the land profitable by bringing in mining companies and settlers.[16] Military forts like Huachuca were established as the first armed force to secure the border territory from roving bands of Apache.

I start with the Huachuca mountain to begin my exploration of the struggle over the control of the visual mapping of land by the encroaching army, surveyor teams, miners, and settlers, who increasingly limited the movement of Apache, Mexicans, and others. Because the power of vision(ing) was entrenched in worldviews, and the control over resources and territory, the Indian Wars mark a critical shift between secular Enlightenment beliefs in rationality, empirically based science, and Euro-Anglo technological innovation, on the one hand, and Native sacred knowledge practices tied to place that stretched from the earth to the sky, on the other hand. Whereas Indigenous knowledge and

Western science are often pitted against each other, locked into binaries of nature/culture, universal/local, and primitive/modern, I contend that Apache tracking intelligence was not prior to but actually foundational to remote technologies that were developed and branded as American military intelligence and surveillance.

While historians tell the story of the Indian Wars as the final triumphant event in the winning of the western frontier, I visited the museums at Fort Huachuca with the goal of questioning why the Indian Wars figured so centrally in the fort's high-tech and intelligence history (and that of Arizona more broadly). I became curious about what constituted the science of military intelligence and how it emerged from Native visual intelligence, especially the Apache scouts' superior ability to track and communicate silently, even to foresee the footprints, or movements, of soldiers and sometimes their own people, who had conflicting ideas about how to confront the onslaught of settlers and the cavalry.[17] I look back to the Apache Wars not only to tell an alternative history about the origins of border security but also to consider what place-based knowledges were suppressed to settle this land, and how Apache sacredscience might offer other possibilities for imagining beyond the "border-biosecurity-industrial complex" (chapter 3), for repairing the land, and for telling a time when the people will return en masse to their land.

The liberal-progressive story of settler colonialism in the southwestern frontier is defined by the "inevitable" evolutionary "reduction" of the primitive by the technologically advanced, or the natural historical evolution in which one species of intelligence is replaced by another, what I call *techno-evolution*. Military historians described Apache communication as code, undecipherable yet reproducible, a natural energy source to be extracted as labor from the bodies and lands of Indians and then bundled back into Western techno-knowledge systems. Within Fort Huachuca's two museums and accompanying archives, the violence of war and bloodshed is naturalized as the progressive motion of historical change that ends with the superior scientific-technological intelligence of Euro-Anglo settlers.

"The Eyes of the Army"

The *Eyes of the Army* statue holds within it a technoevolutionary story first documented in 1877 by Colonel Richard Irving Dodge in his book *The Hunting Grounds of the Great West: A Description of the Plains, Game, and Indians of the Great North American Desert*, then reproduced visually in a photo from the early 1930s (see figure 1.3), then chiseled into a bronze statue in 1995, and then hammered out into a drone in 2019. Situated between the Historical and Intelligence Museums today, this statue visually archives the many layers of historical staging that shape how guests enter the two museums, reinforcing the public's role as witness to Indian "history" as a dying past that continues to be celebrated and *preserved* through novel technological developments, such as the smooth glide of drone "eyes."[18] At the US Army Intelligence Museum at Fort Huachuca, one can visit the Shadow display featuring a class of drone named for its stealthy ability to see and to send data to other locations in the shadow of the visible (see figure 1.2).

Predating the Fort Huachuca statue is a highly staged photo taken at the fort during the early 1930s meant to reenact the iconic scene of an Apache scout dressed for battle as he points two fingers ahead to something beyond the vision of a soldier donning binoculars, or field glasses (figure 1.3). Here Nativision and binoculars vie for dominance, a contest won by the scouts, whom historians describe as the innovative technology that ended the Indian Wars. After the war, when it was no longer necessary to employ Indian scouts, some moved with their families to Fort Huachuca in 1922 to patrol about sixty miles of the fort's boundaries (including preventing Indians and Mexicans from crossing the US-Mexico border). Other duties at the fort included outfitting themselves in traditional dress for the press, for public parades, and for movie directors fascinated with the role of Indians, who were paid to reenact what became famous battle scenes. As was typical in the 1920s and 1930s, the popular iconography confirming the status of the primitive Indian in anthropological studies, newspapers, and films led to them "playing Indian" in traditional outfits with extra feathers, headdresses, and the like.

Despite attempts to assimilate Apache living at the fort, they continued to live in their traditional ways, including living in wickiups, not speaking English, and holding ceremonies. While the *Huachuca Illustrated* defines Apache ceremonies as cultural practices that performed the past ("stirring memories of a proud past"), for the Apache engaging in ceremony was a sacredscience that strengthened their ties to their land.[19] By growing gardens under cottonwood trees where water in this desert region is plentiful, raising livestock, harvesting local plants and nuts, and refusing to live in adobe homes built for them by the army, Apache preserved their worldviews as best as they could despite the occupation of the army, a practice carried on today by those who return annually to the fort. Fort Huachuca's post historian, James P. Finley, dedicated the first three issues of *Huachuca Illustrated* (1993–96) to African Americans (Buffalo Soldiers) and Native Americans (Indian scouts) with funding from the Department of Defense (DOD), which aimed to highlight the cultural diversity of the nation's military history.[20] Historical accounts documented by magazines

Figure 1.2. A Shadow unmanned aerial vehicle at Fort Huachuca's US Army Intelligence Museum, April 2019. Photo by author.

Figure 1.3. Apache Scout William Major with an officer of the Twenty-Fifth Infantry in the 1930s at Fort Huachuca. Source: Finley, "Indian Scouts in Huachuca in the 1920s and 1930s," *Huachuca Illustrated*, vol. 2, 1996, https://net.lib.byu.edu/estu/wwi/comment /huachuca/HI2-25.htm.

and monuments hope to contain Native Americans to the past, erasing the San Carlos Apaches' ongoing land claims on the fort.

The photo, like the Fort Huachuca statue, depicts an Indian scout in traditional dress pointing off into the distance next to a soldier armed with binoculars, a technology that extended vision by narrowing one's overall sight. Even with binoculars, the soldier is unable to see as well as the scout. In 1877 US army colonel Dodge's book-length study of Plains scouts inspired the iconicity of these images, reinforcing how Nativision served as the primitive technology, or code, that led to the winning of the West:

> In communicating at long distances on the plains, their mode of telegraphing is . . . remarkable. Indian Scouts are . . . invaluable, indeed almost indispensable, to the success of important expeditions. The leader, or interpreter, is kept with the commander of the expedition, while the

scouts disappear far in advance or on the flanks. . . . The only really wonderful thing about the telegraphing is the very great distance at which it
can be read by the Indian. I have good "plains eyes"; but . . . even with an
excellent field glass, I could scarcely make out that the distant speck was
a horseman, the Indian by my side would tell me what the distant speck
was saying. Indian signaling and telegraphing are undoubtedly only
modifications and extensions of the sign language heretofore spoken of.[21]

The skills of the Indian scout at Dodge's side inspired a range of military techniques and knowledges, including the ability to exchange secret messages
across long distances ("signaling" and "telegraphing"). In the 1930s photo (and
the 1995 sculpture), it is easy to miss the scout's use of gestural code, or sign language, as he points two fingers out into the distance. For many Native Americans at that time, gestures were superior to language in expressing a world in
constant motion to others across a vast distance. As described by Navajo historian Wally Brown, signing with two fingers off into the distance pointed to
a person (rather than an animal, for instance) by referring to their dual female
and male identity.[22]

Despite settler beliefs that written language expressed white settlers' higher
intelligence, the linguist Jeffrey Davis has documented the many complex
thoughts and ideas expressed through signed language—including a "discussion of the past and future, and distinction of present and non-present
entities—animate, inanimate, human, non-human" as well as "gender and age-
specific activities."[23] Sign language draws from the body's emotive movements
and is place-specific, marking one's position in relation to land—the direction
of the sun, the wind, the mountains, and the action—to make clear how one's
orientation in place shapes what we know. Even sacred relationality could be expressed. For instance, to refer to yourself, you point a thumb toward your heart.
And to refer to a sacred place or being, you direct the thumb from your heart/
self toward the object, to make clear the inseparable relation flowing between
yourself and place, and vice versa, registering the entangled life force pulsing
across all living things. Similarly, to point to some person far off into the distance with two fingers is to honor the shared humanity of the enemy/other,
thus shrinking the distance between Western concepts of self and other.

When Dodge's *The Hunting Grounds of the Great West* was published in 1877,
the study of American Indian sign language was at its height, sending former
military officers to the West to gain insight on how to signal for the US Army
Signal Corps and to learn how to communicate with Native tribes. The rush
to document Native gestural languages, argues Brian Hochman, was a racial

imaginary steeped in the idea of the "vanishing Indian" that participated not only in the urgency to document but in the very kinds of technological innovations necessary to capture a language in motion. As he argues, "the origins of modern media in the United States are distinctly ethnographic."[24] Before ethnography took hold as a discipline, ethnological studies of Native peoples gleaned from mapping expeditions, travel, and wars also inspired novel technological developments in war and border control. For example, Albert J. Myer, head of the US Army Signal Corps, used his knowledge of Native American (as well as Deaf persons') signaling techniques to invent wigwag (aerial telegraphy), a powerful system of long-distance information transmission that gained fame on the battlefields of the Civil War.[25]

Early scholarship by military officers, some recently retired after the Civil War ended, concurred with Charles Darwin's theory of evolution, leading many to consider gesture language as an ancient form of communication by the "Low Tribes of Man" and closer to the communication of animals such as apes, as well as those with disabilities such as the Deaf and even the mentally ill.[26] As such, early scholars of American Indian sign language regarded gestural communication as less evolved than speech and written communication.[27] Yet widespread fascination with Indian sign language, or "gesture code," drove many to study this "primitive" form of communication as an avenue to unlock the mystery of ancient humans.[28] Caught in an evolutionary framework, these writers placed sign language on a more advanced level than pictographic art but lower than the easily archived and universally understood Western alphabet. Ironically, not only did Native tribes such as the Cherokee have a much more complex written alphabet, but various Indigenous peoples saw their visual and communicative technologies as superior to, or at the least on par with, those of white men. This included Iron Hawk of the Sioux, who stated that while the whites were given the power from the Great Spirit to read and write, "He [the Great Spirit] gave us the power to talk with our hands and arms, and send information with the mirror, blanket, and pony far away, and when we meet with Indians who have a different spoken language from ours, we can talk to them in signs."[29] In a similar vein, White Horse of the Arapaho compared the "white man's mechanical medicine" (communication by radio) with the "Indian's medicine" (communication through dreams) to argue that both the white man and the Indian could hear what was not visible.[30]

Ethnologists such as Garrick Mallery refuted their colleagues' belief that Indian gestures were solely a primitive form of communication that predated language. In his address for the American Association for the Advancement of Science in 1881, he stated, "With gesture he could exhibit actions, motions, positions, forms, dimensions, directions, and distances, with their derivations and analogues."[31] Mallery argued that gestures could communicate much more complexity than mere language. Other ethnologists, too, praised the Indians' gestural communication as poetry in motion developed by necessity, such as when hunters found ways to silently communicate with others as they closed in on their prey.[32]

Gestures, like language, communicated worldviews but did so through embodied communication. To express the presence and movement of an animal while hunting, one did so by learning how to enact, or become, this animal. In fact, Gregory Cajete, a Tewa Indian scholar, argues that studying plant medicines and animal behaviors while hunting may be considered the first science. Even though Cajete's assessment of hunting does not account for the feminized scientific knowledge of *curanderas* (healers) and *parteras* (women who aided others with fertility and birthing), he refers to the practical (rather than mystical) biological knowledge that comes from the close study of animal (and plant) habits, anatomy, and needs, including where and why animals move, what they eat (and don't eat), and how to kill swiftly, while preserving the land-animal flourishing for future hunting. Indigenous knowledges are part of a complex place-based cosmology of living and dying, and of learning how to see and know the world from other living beings around them. They learned how to trick animals, which parts of the animal to use, how animals responded, and what they needed to live. As the Apache chief Geronimo himself relates in a translated book, "It required more skill to hunt the deer than any other animal. We never tried to approach a deer except against the wind. Frequently we would spend hours in stealing upon grazing deer. If they were in the open we would crawl long distances on the ground, keeping a weed or brush before us, so that our approach would not be noticed."[33] Through the close study of animals and plants, including eating them, Cajete argues, "we are transformed into each other."[34] To know the self as other was deeply embedded in the liveliness of gesturing, which reflected the motion of life at the heart of their cosmological language. Thus, it is no surprise that Geronimo was a shaman, or medicine man who had a deep knowledge of the spirit, or footprints, of animals and plants and later of the movements of the light-skinned foreigners.[35]

It was the trickster spirit of the coyote that transmitted Geronimo's power as a great warrior. Similarly, for many Plains Indians such as Plenty Coups, communicating with others through their moccasin-telegraph sign language across distance meant embodying "the wolf's power to communicate with fellow pack members across the plains and to camouflage themselves against the plains."[36] As we will see across the book, these tracking and communication skills continue this sacredscientific worldview, even within the militarized ranks of the all-Native American Border Patrol (once called the Shadow Wolves).

It was clear to the Apache that the cavalry soldiers were ignorant not only of the imperatives of warfare but of the place-knowledge all around them. As keenly noted by Leslie Marmon Silko in relation to the chase to hunt down Geronimo:

> The elders used to argue that this was one of the most dangerous qualities of the Europeans: Europeans suffered a sort of blindness to the world. To them, a "rock" was just a "rock" wherever they found it, despite obvious differences in shape, density, color, or the position of the rock relative to all things around it. The Europeans, whether they spoke Spanish or English, could often be heard complaining in frightened tones that the hills and canyons looked the same to them, and they could not remember if the dark volcanic hills in the distance were the same dark hills they'd marched past hours earlier. To whites all Apache warriors looked alike, and no one realized that for a while, there had been three different Apache warriors called Geronimo who ranged across the Sonoran desert south of Tucson.[37]

Military Intelligence: Arthur L. Wagner

Fort Huachuca in Arizona is a military historical site and active national hub for the testing and development of electronic warfare and border-surveillance technologies. Huachuca tests drones, satellites, tethered aerostat radar balloons, communication devices, and other tracking technologies for use along the US-Mexico border as well as for war and peacekeeping in Iraq and Afghanistan. Of the seventy or so military forts erected in Arizona during the settling of the frontier, Huachuca is the only remaining active military garrison. Today the fort educates students in the advanced study of high-tech military intelligence, a curriculum that begins with the Fort Huachuca Historical Museum, built in 1960. Students also come to the US Army Intelligence Museum, built in 1982, directly across from the Indian War Museum, for training and education in military intelligence. Many young people come to the fort to train for jobs and to work in high-tech surveillance for the US Army, the Department of Homeland Security, Intel, Apple, and a range of other industry specialists. The fort's historical legacy in bringing the Indian Wars "to an end" and its strategic location, close to the Mexican border but protected by mountains (perfect for long-range satellite and drone testing), place the fort at the center of military intelligence and innovation and make it a hub for military education.

Created in 1971, the Huachuca Military Intelligence School is one of the largest training centers in US military intelligence. They train "tens of thousands of soldiers, sailors, airmen and marines on an annual basis" by providing courses on intelligence, counterintelligence, interrogation and questioning techniques, weapons intelligence, and much more.[38] In the US Army Intelligence Museum catalog, "The Intelligence Impulse: A Showcase for U.S. Army Intelligence History," one can find officers and their gadgets that revolutionized early mechanized intelligence. According to the catalog, an intelligence officer is characterized by their *intelligence impulse*, or the "inner spark" driving "a vivid but logical imagination."[39] For students of the Intelligence School, the museum "uses history in an attempt to discover what that *intelligence impulse* is" and how, where, and when it originated.[40] As we will see, the intelligence impulse, or the energy driving visionary thinking, springs from the colonial idea of a natural force (Native eyes) driving mechanization. And the pioneers include people such as Lieutenant Arthur L. Wagner.

Wagner, the author of the first US Army text on intelligence, is credited as the founder of US military intelligence education at the turn of the century. He wrote two key texts on security and intelligence that, when published, were authorized as army textbooks: *The Service of Security and Information* (1893) and *Organization and Tactics* (1894). His empirical observations as a military colonel while stationed at Fort Huachuca prepared him to become an intelligence operative who prepared written directives on military techniques to be deployed during the Spanish-American War. In 1898 he was sent to Cuba to set up the first Bureau of Military Information to be organized in the field since the Civil War.[41] Wagner relayed the importance of developing maps, and thus knowledge, of a country before asserting rule over it, a lesson learned during the 1846 US invasion of Mexico. More important, he advocated turning away from the direct-confrontation style of European military tactics staged on the open plains in the East since these methods fail to respond to the style of warfare with Indians on the rugged frontier. US intelligence, he argued, must depart from European strategies of discipline and routine, from rows of soldiers fighting and defending a stable war of position. Rather than rely on formulas borrowed from European battles, Wagner suggested commanders adopt a flexible approach demanded by the uniqueness of each battle, specific enemy, and particular terrain. This homegrown US military intelligence would incorporate reconnaissance, or information based on visual observation of enemy strengths and weaknesses, especially how they navigated the southwestern frontier.

Wagner was not the first to study Apache warrior techniques. At the end of his military service in the northern frontier of Mexico, Don José Cortés wrote a book in 1811—*Memoria sobre las provincias del norte de Nueva España* (*Views from the Apache Frontier: Report on the Northern Provinces of New Spain*)—in which he incorporated his own close observation of Apache warriors, as well as interviews with Spanish soldiers and missionaries who had extensive knowledge about Apache culture. Cortés understood the importance of pacifying the Apache and doing so from a deep understanding of their way of life. He also advocated a simple strategy of deploying Apache scouts to hunt down their own people. Since the colonial entry of Spanish and British subjects, Indian scouts had helped Spanish colonists navigate difficult terrain, learn new food sources, fight in their wars, and identify important landmarks such as water sources. They served as guides for mapping expeditions, especially in demarcating the boundaries of settler colonial property.[42] This practice continued into the 1700s when George Washington, an early surveyor, hired Indians to help him designate the boundaries of his family estate. Indian scouts were even hired to track down fugitive African American slaves, although many more Native American trackers helped slaves escape through routes across Indian territory that were nearly impossible for settler to navigate .[43] Despite this long history of employing scouts based on their intimate knowledge of land, scouts were not officially hired as soldiers in the cavalry until 1886. The Army Reorganization Act of 1866 empowered the president "to enlist and employ in the Territories and Indian country a force of Indians, not to exceed 1,000, to act as scouts, who shall receive the pay and allowances of cavalry soldiers, and be discharged whenever the necessity for their further employment is abated, or at the discretion of the department commander."[44]

Archival sources penned by Spanish conquistadors and the cavalry described how they found the small roving bands of Apache almost impossible to capture through traditional military strategies: "They are not ignorant of the use and power of our arms; they manage their own with dexterity; and they are as good or better horseman than the Spaniards, and having no towns, castles, or temples to defend they may only be attacked in their dispersed and movable Rancherias."[45] Moving the battlefield from the East to the West, from open land cultivated for settlement to "wild" mountainous and desert terrain, reignited a racial imaginary of an impenetrable land and an indomitable people that required an overhaul of strategies and tools of domination. The inscrutable land, filled with curves and crevices hiding a mysterious and wild people, epitomized a natural rhythm that was coordinated, yet impossible to predict, a force of nature that moved to an inexplicable cadence. It was the frontier, and the Apache and Chiricahua warriors' nimble movements, that inspired Wagner to create a unique art and science of war that was distinctly American.

Both Apache and cavalry lieutenants meticulously observed the other. Familiar with the land they constantly moved across, Apache resistance fighters observed the enemy so closely that they could easily ambush the unsuspecting soldiers in their camp. These strong tracking skills allowed them to pursue the surprise attack, a common and sometimes playful tactic when the Apache would sneak up on their enemy and leave a sign that they were there, then steal away without a sound. By leaving behind a footprint, they communicated their superior capabilities and instilled terror into the cavalry. Wagner associated these collective movements with the precision and force of nature: "They fought in successive lines, one advancing when the other retreated; and when they were charged, they scattered only to unite and fight at some point beyond. Their ability to rally quickly often enables them to inflict a heavy blow upon troops disordered by pursuit."[46] Other accounts similarly characterized the Apache as engaging in guerrilla tactics. Without fanfare, or the formalities of "civilized warfare," the Apache "march with no semblance of regularity; individual fancy alone governs. . . . [T]hey move onward indefatigably, with vision as keen as a hawk's tread, as untiring and stealthy as a panther, and ears so sensitive that nothing escapes them."[47] At each turn of the page, Apache movements are invisible to the army, a natural animalistic force, rather than knowledge gleaned in relation to land, including the animals moving stealthily around them. However effective the Apache were, their movements were misunderstood as too irregular to be scientific or premeditated military strategy and thus were regarded as intuitive, or as a force of nature.

The foundations for automated surveillance today can be traced back to the military requisite to not only watch the enemy continuously and closely but to see remotely without being seen. Of first importance was observation, and then concealment, both of which Wagner learned how to accomplish through the Indians, who could conceal themselves by blending into the environment.[48] This late nineteenth-century language echoes drone warfare today. Armed with ever more advanced cameras, the drones on display at the military museum can track and record enemy movements from high above without being seen from below. Yet, rather than see from above, Apache and Sioux scouts were able to maneuver undetected on the ground.[49] Wagner and other lieutenants studied these scouts as idealized soldiers since "nothing escaped their notice; tracks, broken branches, upturned stones, ashes of camp-fires, horse-dung—everything was noted and reported back to the commander."[50] Countless accounts described the Apache, Sioux, and other tribes as melting into the mountains, trees, or rocks, or using the terrain to hide and move stealthily, perfecting the military tactic of moving without detection and accomplishing the surprise attack.

Wagner begins his chapter "Outposts" with this quote: "To exercise ceaseless vigilance, to be in constant readiness for action, and to preserve the most profound silence are the cardinal principles of outpost duty."[51] How the scouts moved undetected is worth quoting in detail:

> The scout gains some high point, where, lying on his belly in the shadow of some tree or rock, he sees everything without being seen himself. . . . The expedients adopted for concealment are many and ingenious. The scout sometimes crawls towards a rock on the crest of a hill, and when near it draws his blanket, or a white cloth or stable frock (according to the color of the rock), over his head and shoulders, covering everything but his eyes, and then wriggles himself by degrees up to the rock, where he remains motionless until he has minutely scanned all the country in sight, when he withdraws as stealthily as he approached, whether anything has been discovered or not. He often conceals himself by holding a piece of sage-brush in front of him while lying down. Sometimes he fastens bushes to the upper part of his body, extending above his head; then sitting in a "wash-out" wallow, he is completely concealed, while his own view is unobstructed.[52]

In Wagner's account, his own military reconnaissance takes seriously Native people's skills as worthy of deep study and emulation. Wagner's archival evidence, based on empirical documentation, is limited, however, by Western ways of knowing. Clearly, this depiction of the Indian scout is filtered through a military-colonial-settler imaginary that is well trodden in a range of colonial films and popular representations of the savage Indian with tribal markings tiptoeing across the landscape hidden beneath a leafy bush. These depictions downplay Indigenous intelligence through a popular conception of the noble savage in harmony with nature, while also perpetuating the association of Indigenous peoples with clever deception (chapter 3). While Native people were no doubt good trackers, these archival remnants reinforce empirical observation as the real, or objective, evidence of historical accuracy, further estranging the possibility that Nativision entailed much more than "meets the eye."

For Wagner, innovative methods, based on enemy reconnaissance, reiterated the need for empirical evidence and professional theorizing rather than drill and administration, for problem solving rather than routine.[53] An entire automated social world was emerging, one inspired by the need to incorporate the swift and defeating blows of the Apaches' natural skills and configurations, yet limited to the confines of Western belief systems. Settler faith in a distinctly American character of innovation that could tame the chaos of the frontier and its inhabitants was premised on the superiority of human technological intelligence over the natural world. While the Spanish and the Mexicans were unable to control the Apache, the Euro-American military prevailed, but only when they figured out how to adopt Indian scouts, and the force of nature more broadly, as part of the repertoire of innovative technology, or as "the eyes of the army."

In the section "Organization and Discipline" in Wagner's book *Organization and Tactics* (1894), his discussion of military discipline foreshadows the promise of today's automated technologies:

> A perfect army would be one in which each part could respond to the will of the commander as quickly and certainly as the muscles of the body respond to the *impulse* of the brain. The more closely a military force approaches to this impossible ideal, the more does it merit the title of an army; and the farther it recedes from it, the more certainly does it become a mere *armed mob*, highly susceptible to the influence of chance, and uncertain in its action, even when opposed by a foe no better than itself. . . . The entire theory of organization rests upon the principle of individual responsibility and subordination, so that, no matter how small or how great the number of individuals gathered together, some one is responsible, to whom the others must be subordinate. This responsibility and subordination are the great factors in the control of the army.[54]

Wagner's sense of the importance of military hierarchy was influenced by German military education. Without hierarchy and the subordination of the individual will to the greater good, Wagner saw chaos. At the helm was the brain that commanded not only the arms and legs of the army but also the individual soul of each man, who was simply a part of the collective body. To capture this impulse, the commander must appeal to the *individual* character that inspires his men. While the French were moved by glory, and the British by a sense of duty, the American solider could be appealed to through "common sense, pride, and patriotism."[55] Although individualism was praised on the frontier, this trait must be suppressed through discipline. The seeming contradiction between intelligence and discipline, democracy and authoritative control, speaks to early fears of social disorder given the mass movement of immigrants, freed slaves, and Native peoples after the Civil War. Mechanized labor and technological innovations contributed to these shifts and also promised to alleviate chaos, or the unknown effects of rapid social and technological change. Even emerging notions of science were based on the need to obliterate chance, the unknowable (chaos), for a present-future that could be predictable and certain.

Apache warriors and nature emerged as unknowable threats in contrast to this predictable environment. The temporal flow of Apache surprise attacks, even when violent, harmonized with their surroundings, what Wagner described as a flow of lava silently suffocating the army. At the same time, the temporal flow and spatial command of Apache movements communicated a belonging to place that contrasted with the foreign mechanical militarized tempo of settlers and soldiers. One of the few commanders able to learn from the Native Americans he met all across the United States was General George Crook. During his command in Oregon and California, he learned navigational and fighting techniques and even how to communicate with the Apache, some of whom gave Crook the name Nantan Lupan, or Grey Wolf. He, too, recognized the superior skills of Natives on their land. He said, "It would be practically impossible with white soldiers to subdue the Chiricahuas in their own haunts. The tendency of military drill discipline is to make the military soldier a machine. He cannot compete with an enemy whose individuality is perfect."[56] Euro-American military discipline and industrialized capitalism are perceptively noted by Crook as a disadvantaged bodily discipline that dulls the mind and drowns out the spirit of individualism so coveted in the "new world." In fact, Native American individualism formed the emerging character of an American spirit set apart from their European counterparts, especially in the face of the keen and perceptive skills of nimble Apache who moved like the wind and could disappear from view and reappear for a lethal surprise attack. Apache descendants remembered stories confirming the foreign tempo of the soldiers: "My dad says the cavalry would come here, and the Apaches would watch their drills and laugh 'cause you could never get an Apache to fight like that."[57] The idea that one would stand still and shoot was ridiculous as it left the soldiers frozen in space, a sitting target easy to attack.

Intelligence gathering could be done through reconnaissance by military officers, scouts, spies, and slaves. Wagner devotes an entire chapter of *The Service of Security and Information* to women spies, and especially Black female slaves, who proved ideal intelligence gatherers as they could move without being detected—or seen. Black female slaves (such as Harriet Tubman) were invisible, he argues, because they were unthinkable as intelligent or educated, brave, or even physically capable of military endeavors. And they were thought to lack character.[58] Native women, while not addressed, were also well-known symbols of peacekeeping as they were forced to move between tribes or even between Euro-Anglos and various tribes as the medium for communication, for sharing of gifts, and for trade.[59]

While much can be gleaned from Wagner's account about the relationship between warriors' skills and Indigenous hunting and planting practices and familiarity with the terrain, he does not consider the making of the warrior within the settler colonial context of violence and dispossession of land, as well as the cosmic worldviews shaping the production of the warrior.

Apache sacredsciences, or deep observational practices meant to preserve communication and sacred relation with ancestral land, were interpreted by military officials such as Wagner as part of their natural aptitude for war. Wagner naturalizes this warrior spirit as a deep cultural practice disconnected from the context of settler colonial violence:

> It should be observed that these Indians are all trained to war, and that their methods are not the result of the inspiration of the occasion, but of constant practice, and of a study which is not less deep because it is unlettered. Methods of scouting, various expedients of warfare, and even geographical details, are learned by one generation from another; and more than one instance has been known of an Indian finding his way without difficulty thru a country which he was traversing for the first time, because he had learned so thoroughly from others the relative positions of prominent landmarks as to be in possession of a reliable mental map. Constant practice in hunting, stalking game, and making long journeys thru wild country makes the Indians expert in judging distances, reconnoitering, utilizing cover, and husbanding the strength of themselves and their horses.[60]

Although popularly known as the fiercest Apache warrior in US history, Goyathlay, or Geronimo, grew up on the top of a mountain, close to the visioning power of the stars. He was raised as a powerful medicine man, a shaman, and a skilled hunter. In the documentary *Geronimo and the Apache Resistance*, some of his descendants remind viewers that he became a warrior only after Mexican troops stormed into his community and slaughtered his wife, mother, and children right in front of him. It was then that he became a warrior, when "his heart ached with revenge."[61]

According to a fellow Apache interviewee, Geronimo was a shaman who "was said to have the 'power of the coyote,' as he could appear here, then suddenly over there. He could outfox anybody. The soldiers passed right by him."[62] It was said that even their bullets were no match as Geronimo blessed the war equipment so a bullet would not hurt him. What Wagner and others narrated as the ability of Indians to move silently and without detection through the land was understood by Chiricahua and Apache descendants as a shaman's ability to take on the spirit of an animal or other natural forces. Shamans could move beyond the limit of human vision by transforming themselves into the form of an animal or a bush, or a slice of shade angled between two uneven jutting rocks at a particular time of day. A famous warrior woman, Lozen, fought alongside Geronimo and shared some of his superhuman capabilities. It was said that when she prayed, her outstretched "hands tingled and her palms changed color" to indicate the location of the enemy.[63]

Indian scouts hailed from many tribes and served the army for a variety of reasons. Some hoped they could mitigate the death and damage done to their people by the US Army. Others sought financial compensation, status, and revenge against raiding Apache. Still others saw the numbers of Anglo settlers increase to such a level that they believed their future could be sustained only through collaboration with whites. There were whites such as General Crook who earned the highest respect from the Apache. Crook's deep understanding of and communication with the Apache allowed him to hire the largest brigade of Apache scouts in military history to capture Geronimo and his small band. Many Apache wanted a swift end to the bloody massacre of their people by the army and settlers, especially as deaths swelled owing to starvation. Hungry, tired, and desperate for peace, many agreed to help Crook bring down the last group of resisters, the last band who refused to give up their expansive land and hunting practices for a settled life of raising scant cattle on desolate reservation land. Crook gives full credit for this victory to the Indian scouts. He noted that from 1872 to 1874, "of the 283 hostiles killed, 272 were accounted for by Indian scouts, while they captured 213. In contrast, commands without scouts captured and killed (together) fewer than 20."[64]

As mentioned earlier, even during periods of peace, many scouts were kept on the army's payroll. Some monitored the borders of reservations to prevent escape or to track and return runaways to the reservation. Once Geronimo surrendered in 1890, the need for Indian scouts was greatly reduced, and their numbers dropped from 150 to 50. Those retained assisted the army during the Spanish-American War, then aided General John Pershing in the Mexican war against Pancho Villa in 1916 and served as a top-secret force called the Alamo scouts during World War II, shaping the reconnaissance mission of today's Border Patrol and Secret Forces. Some Alamo scouts (including a few from the Lakota Sioux and Chippewa tribes) served with the Navajo code (or wind) talkers. Most Alamo scouts, however, performed reconnaissance and raiding across enemy lines, sometimes dressing as the enemy in order to gain valuable information, to recover American prisoners of war, and even to steal away Japanese who became prisoners of war on the US side. Even before the war, Indian scouts served as the most skilled trackers patrolling the border between the newly won southwestern territory and Mexico.

Not all lieutenants, however, were convinced of the superior skills of Native peoples. After Geronimo escaped for years from an army that far outnumbered his band and then from the reservation a second time, Crook stepped down and was replaced by Nelson M. Miles, whose arrogance was later deflated. In the documentary *Geronimo*, General Miles is quoted as saying, "With our superior intelligence and modern appliances, we should be able to surpass all the advantages possessed by the savages."[65] Ironically, even with five hundred troops, more than any previous brigade, Miles's mission failed as they found the "mountain being almost impassible." The voiceover in the film *Geronimo* is accompanied by the image of a tall mountain stretching above a thick cloud mass as Native singing and drumming fill the scene with the sense that the mountain, responding to the songs of prayer, prevented their passage, turning the goal to hunt down Geronimo into a military failure. After six months, Miles finally had to adopt Crook's method, and his success in finding Geronimo this time resulted from two Apache scouts, Geronimo's own relatives. Word spread that the elderly folks were angry with Geronimo for causing so many of their people to die. Based on their pleas for surrender, Geronimo and his band were willingly found. The documentary *Geronimo* decenters the human arrogance of white army officials, whose power wanes in relation to Native ancestors, especially when faced with the powerful spirit of Huachuca mountain.

In his 1896 memoir, Miles describes the Apache as savage "outlaws to be tracked and subdued," especially in the places where they evade detection—the mountains.[66] He states, "The mountain labyrinths of the Apache may be compared to the dens and slums of London, though on an immensely greater scale . . . for cunning, strength and ferocity have never been surpassed in the annals of either savage or civilized crime."[67] He touts his cunning ability to bring down Geronimo as an inevitable outcome of the superior technology of the heliograph mirror.[68] This wireless telegraph system predated the electric telegraph and was set up on high mountain ranges where little atmospheric moisture or disturbance allowed them to send communication signals by flashes of sunlight reflected by a mirror. Under Miles's direction, the US Army Signal Corps established fourteen heliograph stations across Arizona, including one at Mount Graham in Tucson, Arizona, that was one hundred and thirty-three miles away. On this same mountain, one hundred years later, the Apache are still fighting to reclaim their sacred mountain, where the University of Arizona built a large observatory complex with thirteen telescopes armed with multi-mirror optical lenses. As we will see in chapter 3, these heliograph stations were critical technological hubs where the cavalry communicated the location of "savage criminality" across the vast and wild land of the Apache. A generation later, these mountaintops and the use of light flashed across space have inspired the telescopes erected to see the unknown terrain of outer space. Sensors designed at the University of Arizona today also apprehend the unknown terrain of the inner body's errant physiological movements at global borders around the world.

Return of the Repressed

Soldiers at Fort Huachuca today continue to learn valuable lessons from the Indian Wars, such as the tactics used by insurgents in Afghanistan who camouflage themselves (or take cover) by blending into trees, boulders, and houses. Just as the Apache became visible to cavalry soldiers only moments before they raided or attacked, today US soldiers in the Middle East or at the US-Mexico border say they feel that they are fighting an "invisible enemy."[69] These techniques are constantly updated in a dance between new innovations in militarized techno-vision from above and disappearing acts by those on the ground. For instance, Taliban fighters become invisible by tricking the heat sensors of the Apache helicopters and drones. They transform themselves into nonhuman forms by "covering themselves in a blanket on a warm rock."[70] One sophisticated mode of seeing meets another savvy technique for disappearing. While battle on open ground and in airspace is an occasion for swift and deadly warfare, the need for better intelligence on the ground has led to an intelligence program in Afghanistan called "Human Terrain." Trained at Fort Huachuca, troops learn not only how to maneuver (and develop technologies and knowledge) in the "physical terrain" of war but also how to manage its "human" dimensions, or "local people's viewpoints and practices, allegedly acting as cultural interpreter."[71]

After 9/11, Fort Huachuca welcomed a surge of resources from the government owing to its location in a desert landscape similar to that of Iraq and Afghanistan, providing the most realistic military training found in the United States.[72] In Congressman Ike Skelton's well-circulated article published a month after 9/11, "America's Frontier Wars: Lessons for Asymmetric Conflicts," the Indian Wars continue to provide the historical example of why military strategists should not downplay the power of the subordinate foe. Similar to today's guerrilla fighters in the Middle East, Indian guerrilla tactics return time and again as a formidable threat to the dominant powers of US military technologies. While the US public expects precision warfare, from the US military in regions such as in Iraq, to be bloodless and short-lived, Congressman Skelton warns that future asymmetric conflicts will be more dynamic and lethal, "marked by greater intensity, operational tempo, uncertainty and psychological impact."[73] Opponents today are said to use deception effectively, just like Native peoples' superior knowledge and use of the rugged terrain of the West. He recounts the story of "Indian skulking tactics" of concealment and surprise. In this story, a band of Indians concealed their tracks by stepping into each other's footprints, fooling troops who (mis)read the number of Indians ahead and thus foolishly entered into an ambush in which most of the soldiers were killed. US forces today, Skelton states, will face similar uncertainty "as adversaries mask the size, location, disposition and intentions of their forces."[74] In addition, those with superior knowledge of the terrain "will seek to offset our air, intelligence, surveillance, reconnaissance and other technological advantages by fighting during periods of reduced visibility and in complex terrain and urban environments where they can gain sanctuary from US strikes."[75] And similar to Native Americans who were armed by Europeans with weapons superior to those of the colonists, military strategists today worry that the misperception of insurgents as "primitive" people will continue to hack even the most advanced technological approaches. Skelton's article turns back to the Apache, who learned quickly how to dismantle army communications systems during the 1800s: "The telegraph line, which once had commanded their awe, no longer was mysterious. By 1882, the Apache had learned its function and its method of operation. When they jumped the reservation, they would cut the lines and remove long sections of wire, or they would remove a short piece of wire and replace it with a thin strip of rawhide, so cleverly splicing the two together that the line would appear intact and the location of the break could take days of careful checking to discover."[76]

The anxiety here is that when a superpower grows too large and is blinded by an arrogant sense of its dominance, the smaller, nimbler guerrilla forces will enter the system like a viral threat that destroys the heart of the empire from within. And what happens when the memory of Apache peoples returns back to the land? What happens when the tracks of the past held onto by the mountains, wind, rocks, and ground are heard by the people as a plea to return? By the end of the nineteenth century, the Apache had lost six million acres of land. Despite this loss, Apache persist in maintaining ties with their homelands, which have been occupied since the Indian Wars. And the mountain does not forget. It will continue to hold the memory of them until the people return.

Conclusion

This chapter takes us along an itinerant path back to the Indian Wars as a way to remap Indigeneity as the founding logic for threat on the border today. Historians retell this history through a liberal framework of technological uplift that aligns with attempts by the military to assimilate Nativision into the very military optics of automated warfare. By tracking the absent presence of Indigeneity within historical narratives and the technological destiny of US empire, we can see how historical (narrative) and material (techno-objects) accounts enclose Native presence in a primitive past destined to be replaced by the more civilized settlers.

At the same time, the military's assimilation of Native code into an armament of border control, war, and empire is under constant threat of unraveling. As Skelton argues, the frontier and the Indian reappear to warn military strategists not to downplay the sophisticated tactics of Indian guerrillas. Native intelligence provokes a psychic unsettling that continues to haunt the West as military strategists acknowledge that Native warrior techniques persist in insurgent strategies that threaten to dismantle the technological might of the West. In addition to considering Indianness as a figuration of the Western colonial imagination, I also want to consider Indigeneity as a practice of making the land into a powerful animate relation. Apache sacredscience continues to threaten the logic and power of Western sovereignty and borders. Their relation with Huachuca sets in motion the return of one generation after another to the fort. This quiet return to Huachuca by the Apache is connected to a global movement of Indigenous land-based resurgence. As we will see in chapter 2, the struggle over how to "see" the border, especially on the Tohono O'odham reservation, continues to be fraught with contestations over border boundaries and fences, sovereignty, and the practices that maintain sacred relations with land.

TWO

Occupation on Sacred Land *Colliding Sovereignties on the Tohono O'odham Reservation*

Trouble on the Tohono O'odham reservation reached a particularly high pitch during the 1970s when the media ran a series of articles fomenting public fear about a security "void" on the reservation, what journalists called a "borderless border" and a black hole.[1] In this chapter I ask, What happens to our understandings of borders and belonging when we consider sovereign nations such as the Tohono O'odham, whose reservation is characterized as an empty space, a lawless zone or dark hole outside the national security gaze? Hypermediated exposure to the border as a "wild," empty region outside state law, access, and control reconfigures the border as a colonial site/sight and rugged frontier, a shadowy terrain beyond law, Western access, vision, and control. It has also been perpetuated more recently as an unpatrolled desert death trap for migrants. Media and scholarship abet federal and state funding when they characterize this desert region as an unsafe place where sophisticated technologies are needed to "see" and secure the border. Two congressional reports appealed to this logic of the void by stating that drones along the border would "eliminate surveillance gaps."[2] From the 1970s to the present, the Tohono O'odham reservation's supposed borderless status, and its tenuous sovereignty at the border between two nation-states, justifies its occupation by the Department of Homeland Security (DHS). Today O'odham tribal members refuse the imposition of walls, security towers, drones, Border Patrol agents, and at least nine other US security forces to control the crossing of migrants and drug runners across their land and into the United States. The intrusion of military security into the O'odham reservation rehashes the making of this land as a wild and sparsely populated frontier in need of conversion into a securitized border. Ironically, in 1978, the Supreme Court decided that Native tribes do not have criminal jurisdiction over non-Indians on their reservations.[3] Thus military occupation by DHS border security continues a long colonial history of dispossession of O'odham land and the attempt to undermine Indigenous sovereignty.

The colonial frontier meets the border through depictions of this desert region as lawless, as at the edge of jurisdictional and sovereign reach, or as primitive and underdeveloped. Lauren Benton complicates imperial borderlands as "zones of legal anomaly—produced by conditions of contested and multiple legal authority—rather than as zones of lawlessness."[4] What is happening on the O'odham reservation resonates with other Indigenous communities, such as the Yaqui and Kickapoo on the US-Mexico border or even the Iroquois peoples, whose land straddles the US-Canada border. Audra Simpson, a Mohawk anthropologist, reverses the logic of the media and laws that criminalize the Mohawk as "people without law, as people who transgress borders, rather than refuse them *lawfully*."[5] O'odham activists also refuse border security *lawfully*, as a force of occupation and assault against their autonomy as a sovereign people. That said, refusing state security has led to their characterization as being against the "public interest" and thus as threatening the nation, as lacking patriotism, and thus potentially not deserving their much-needed state funding. Yet some O'odham members reverse the logic of threat aimed at migrants and their own community and tell another story that emphasizes the state's role in perpetuating the conditions for militarized occupation of their land.

Given the shift in border control from arresting migrants in the interior to controlling borders at busy ports of entry since Operation Gatekeeper in 1994 (when walls and fences were built along the border from San Diego, California, to Yuma, Arizona) and Hold the Line in 1993 (when walls were built from El Paso, Texas, to Arizona), US federal laws have funneled border passage into more desolate and dangerous desert terrain.[6] These policies brought border traffic directly onto the Tohono O'odham land, which occupies a sixty-two-mile strip of land (as the eagle flies) that straddles bordered land between Arizona and Sonora, Mexico.[7] Some on the reservation see this as intentional. And they are not alone. There are over forty-five Indigenous nations whose territory spans the US border with Mexico or with Canada.[8]

The depiction of the border as a void erases longer histories of settler theft of land considered wild and empty, or void of inhabitants.[9] It is the government that has produced the O'odham reservation as a testing ground for surveillance, a space that must be cleared of all intrusions—migrants and O'odham tribal sovereignty—to create a surveillant zone free of *interferences* that might block remote communication by the border-security apparatus to Border Patrol agents. With checkpoints across the reservation, everyday movement for the O'odham has been converted from a sacred aspect of their collective sovereignty into a protected right for individual tribal members who can prove authorization to be on this land.

Against the state's sovereign control of mobility as a right, O'odham consider movement across their land as a sacred practice tied to their autonomy as a people, a responsibility embedded within their way of life, or Him'dag. O'odham elder Ofelia Rivas defines Him'dag as "the canon of beliefs, stories, and rituals that governs O'odham life" across all O'odham territory: "from the northern reaches of the desert to the Cortés seashore, is holy land. We are directed by Creation to maintain the area by doing our ceremonies. By doing our prayer offerings. Doing our songs to specific mountains. Gathering medicine."[10] Some also define Him'dag as "Path," and as a verb it means "to be able to walk."[11] I humbly learned a little bit about the importance of O'odham worldviews during my visit to the reservation in March 2019.[12] As I understand it, in O'odham cosmology, the symbol of footprints on a sacred path tells an important origin story guided by the ancestors who came before, while also marking the path for those yet to come. Movement is at the core of life and thus must be respected. Rivers must flow, rain must have good earth to seep into, and people's pathways to cultivate these relational bonds must be honored. The concept and material construction of national borders violently defy, or block, these flows of life.

We were invited to visit the Tohono O'odham reservation after Iriany, whom I now consider kin, welcomed her O'odham family and friends to dinner at my house. We discussed the myriad hardships faced by the O'odham nation owing to the presence and occupation of border control on their land. I explained my project and realized how important it was to share their experience of the border for this book. They implored that I visit their homeland before writing about it. When we arrived, we were driven to some of the sacred sites where stories are held about the origins of the O'odham as a people who have lived on this land since long before the Spanish, Mexican, and Anglo colonists arrived. After our O'odham friend obtained permission from the tribal council to drive us to the border, the Border Patrol watched us suspiciously as we drove along the border fence, which many on the reservation see as an affront to their sovereign right to move across their land, which spans far beyond this artificial line. O'odham are treated as illegal immigrants on their own land. In addition to stopping migrants, border patrol officers detain tribal members at border checkpoints, where they are harassed for tribal cards or proof of US citizenship to show their right to be on their own land.

As the second-longest border zone on a Native reservation, it is surpassed only by the US-Canada border zone in the Iroquois tribal nation. As argued by Mohawk scholar Audra Simpson in the case of the Iroquois peoples, crossing borders is less an "occasion for transgression, a means of decentering the national narrative of a culturally homogenous and monolithic nation-state" or of moving through "juridical identities."[13] Instead, crossing borders is a key component of articulating their autonomy as a sovereign nation with the ability to determine how they move across their land. In fact, their way of life, or Him'dag, is inseparable from their sacred right to movement across land, an O'odham expression of a different relation to citizenship, or belonging. When the state offered US citizenship to all O'odham members, most refused this "gift" as another attempt to replace their sovereignty with that of the state. State confirmation of citizenship is not an inherent right naturally available to all but an exclusionary and violent ideology and practice that normalizes rights, protection, and life for some while regarding others as undeserving.[14]

For the O'odham, sovereignty goes against having to request individual rights authorized by the state. Instead, they articulate their internationally recognized right to collective self-determination, or the ability to make autonomous decisions as a sovereign nation. They demand the right to engage in land-based practices and actions that materialize the sacred flourishing of life for all inhabitants. They understand all too well that, as argued by Andrea Smith, "the goal of colonialism is not just to kill colonized peoples, but to destroy their sense of being people."[15]

O'odham land has stretched across Arizona and Sonora, Mexico, since time immemorial, long before the concepts of citizenship, the nation-state, and borders were developed. In fact, when O'odham members such as Verlon M. Jose and Jacob Serapo say there is no word in their language for "border," "wall," or "citizenship," they understand the material force of words to bracket the world, amputating their bodies, land, and culture.[16] Words brought onto their land by outsiders, and especially by the Department of Homeland Security (DHS), reinforce an alien occupation aided by technologies such as treaties, maps, a fence, the Border Patrol, surveillance, and especially a future border wall. Many reject state checkpoints and harassment by border agents who demand they declare themselves to be US citizens, as this would continue to authorize the state to be the arbiter of rights and territorial membership and exclusion, thus stripping them of their sovereignty. As Nicholas De Genova argues, "The legitimacy of modern state power originates from a mythical covenant, a 'social contract,' among naturally free and equal individuals. . . . Citizenship, then, is necessary to translate this wild, 'natural' freedom into the sort of political and juridical liberty that can be used to justify the authority of the state as the 'democratic' expression of a popular will."[17] The myth that we each hold an "'inalienable' birthright freedom in return for 'rights' granted by the state" is multiply violent.[18] It reinforces the belief in individual freedom and the notion that this freedom is limited to a human body with clear geopolitical boundaries. Ironically, these individual and inherent freedoms must be protected by the state and law. Here, too, belonging established through documentation such as birth certificates—rather than through relation with land—reinforces static relations to land-as-property.

Thus, this chapter considers the effects of militarized border security not on border crossers but on those whose lives have become bordered.[19] The phrase *to be bordered* is my way of suspending our assumption of this land as already a geographic border between nation-states and asking how the DHS attempts to *convert* O'odham sovereign land into a border zone. Turning *the border* into a verb forces us to think about this land as contested terrain between the state's colonial imaginary of a frontier in need of control and tribal members' struggles to maintain their identity as O'odham, rather than as either US citizen or undocumented migrant.

The majority of scholarship and activism on immigration and border studies continues to absent Indigenous peoples from a static borderland that is merely crossed by migrants.[20] Only recently have critical debates emerged between Latinx migration studies and Indigenous studies to bring attention to the collisions between settler colonialism and immigration, especially how state violence tangles the lives of migrants, many of whom (even other Indigenous migrants) are channeled onto seized Native land. Unfortunately, migrants' presence contributes to the settler colonial project of Native American dispossession.[21] In this case, migrants and drug runners cross their land, rather than settle on it, creating a funnel to the United States.

Rather than blame migrants (although some do), many O'odham describe the state's enforcement of a border on their land as a disorienting state of living that threatens to render them extinct. Thus, part of what I explore here is that the experience of borders is not universal. O'odham members' sense of disorientation complicates Gloria Anzaldúa's theorization of the borderlands as a *choque*, or a collision between opposing worldviews that forces one into a state of *nepantla*, or of being in between worlds, a dismemberment of self and community that, however traumatic, may also lead to creative alien ontologies.[22] While the clash of sovereignty on O'odham bordered land emboldens O'odham claims to ancient knowledges and practices, they are less invested in decolonizing their expulsion from Indigeneity, or from the human. In other words, the techno-gaze of the state—rather than functioning as a violent imposition that leads to novel ontoepistemologies, of being and knowing oneself otherwise—prevents them from *persisting* as O'odham. Instead, they fight against the occupation and carceral containment of their land by a militarized apparatus that uses virtual and material techniques (surveillance towers, the Border Patrol, a border wall, etc.) to disorient and alienate them from land-based knowledges and practices and thus crush their ability to exist.

That said, O'odham tribal members are subjected to power in ways that proximate the experience of border crossers, but for the O'odham, this happens on their own land, transforming their home into a strange and dangerous place. According to the website *O'odham Solidarity Project*, surveillance towers, the presence of the Border Patrol, and a wall only work to divide their people, scar the earth, halt long migration practices, and compromise tribal sovereignty.[23] In scholarly analysis of power, especially through the one who rules, most theorizations of sovereignty focus on anthropocentric power, or the sovereign's right to rule (oftentimes wielded through violent means) over land, territory, and "life itself." Surveillance is a system that seeps across and weaponizes the environment, from the land to the sky, creating what Kristen Simmons calls *settler atmospherics*.[24] While she employs this term with respect to the #NoDAPL movement, there is an everyday feel of warfare on the O'odham reservation, where helicopters, drones, the Border Patrol, and ground and air sensors occupy the reservation from the sky to the land.[25]

We miss key aspects of O'odham sovereignty when we fail to consider their sacred responsibilities to protect the movement of humans, animals, and even water and wind across land, including unobstructed access for songs, prayers, and words to find their path through the air and sky, as I discuss later. This has become a particularly thorny issue as the O'odham, and other bordered reservations, are occupied by drones, helicopters, surveillance cameras, and other technologies that invade reservations without always entering onto, or touching, land. How might we think critically about the sovereignty of all life from above as much as from below?

The production of the border is highly mediated and entangled in multiple scales and contested notions of sovereignty. Drones and the Border Patrol surveil and capture not only migrants and drug runners but increasingly local O'odham peoples. Thus, I assess the complexity of border security and surveillance on the Tohono O'odham reservation from the perspective of media accounts in both Native and US news sources, US government reports on the border, and O'odham residents and activists.

Bordered Lives and the Promise of Citizenship, from the Gadsden Purchase to "The Wall"

If we look back to nineteenth-century mapping surveys of the US-Mexico border, the US Corps of Topographical Engineers (made up of an elite class of West Point graduates) created the first definitive map of the border after the Gadsden Purchase of 1853. This expedition brought together a motley band of surveyors, biologists, zoologists, soldiers, artists, geologists, and Indian scouts who risked their lives to collect extensive, and potentially lucrative, information about the border region.[26] The volumes of information gathered reinforced colonial expansion and wealth in numerous ways, from archaeological, botanical, and zoological collections sold to museums (such as the Smithsonian) to lucrative German artwork that exposed this "unknown land" and people to a curious public. Indian scouts or guides, many of whom were women, were critical to the safety and success of the expedition teams, serving as translators, peacemakers, experts on plants and water sources, and savvy navigators of diverse regions.[27] Also significant was the use of maps and data to document safe versus dangerous zones of the border region. When they encountered Mexican and Native villages that were evacuated of life, they read this absence through the lens of Indian threat, or as proof of Indian raiding zones. That many left their homes seasonally, depending on the weather and shifting bounty of the land, or for pilgrimage, fell outside colonial perception.

Surveyors secured state sovereignty through their depictions of border life, especially by directing where to draw the new border line. This information facilitated state directives as to where to safely situate settler colonies, where to build roads and railways, and where military garrisons needed to be placed, as a tool of warfare, to control the movement of Apaches, who were known to raid from on both sides of the border. These survey teams reinforced the idea of a terra nullius where signs of life were stripped from the scene. This documentation of absent presence—of scant animals, scattered plants, evacuated homes and mining compounds, and sparse water locations—also devalued the land the United States would purchase from Mexico.

The demarcation of a national border did not always affect the O'odham (whom the Spaniards, and then other Europeans, called the Papago), who lived fairly autonomously until the 1853 Gadsden Purchase, when the United States bought a landmass from Mexico that later became part of New Mexico and Arizona. This binational agreement between the Mexican and US governments allowed the United States to seize thirty thousand square miles of land in Sonora, Mexico, including O'odham territory. In addition, the United States imposed a border in the middle of the Tohono O'odham territory, severing the land in two. When the Treaty of Guadalupe Hidalgo was signed in 1848, most O'odham found themselves on the Mexican side of the border. Yet in 1853, when the United States paid Mexico $10 million for land that included southern Arizona for the construction of a transcontinental rail line, the border was redrawn again, and O'odham land shifted again, so that most of their territory lay on the US side.[28] In addition, this agreement included a provision by Mexico that the United States could buy this territory on the condition that they would prevent "marauding Indians" from crossing into Mexican territory.[29] Thus, military outposts were set up along the border starting in the early 1860s to control the movement of Native peoples, especially the Chiricahua Apache.

Missed in many historical accounts of the Gadsden Purchase were the myriad ways Mexican and US developers and farmers destroyed O'odham homes as they seized land and cattle that had been left behind *temporarily*. Some O'odham migrated to work in seasonal labor markets, or for seasonal stints after droughts, moving onto more fertile land with better food sources. At other times, they would leave home to participate in ceremonial festivities. The Southern Homestead Act of 1866 legalized this theft by offering their land free of charge to white citizens who wished to farm the land, including women and immigrants who applied for US citizenship. The privatization of land was encouraged as a strategy to spur economic development and to redistribute land owned by large slaveholders but also supported settler theft of land that was supposedly empty, abandoned, or unused.[30] When the O'odham returned to their homes, they had little recourse when US or Mexican officials demanded proof of landownership or citizenship status. Thus, in 1872 congressional discussions emerged declaring the need to create a demarcated land base with documented boundaries and better governmental oversight (surveillance) to prevent squatters, profiteers, and others from encroaching further onto Native lands. This seemingly benevolent gesture coincided with the time when various reservations were set up in the Southwest to contain and monitor

Native peoples, who faced capture by Indian scouts or military personnel if they attempted to leave. Philip Deloria describes these historical practices of surveillance on the reservation as "a colonial dream" where "fixity, control, visibility, productivity, and most importantly, docility" could be realized.[31] In response to the great loss of land during the second half of the nineteenth century, President Woodrow Wilson established a permanent reservation for the O'odham in 1916 and erected the first US-Mexico border fence there. The fence was built to protect O'odham land, but US soldiers also protected this new international line to prevent Mexican revolutionaries from entering the United States, as well as to sediment the practice and idea of a territorial border. In doing so, the federal government continued its colonial practice of occupying Indigenous land through projects aimed at securitizing the border. In fact, when the US government creates Indigenous reservations, the title is held in trust by the federal government, making it easy for the state to expropriate land for federal projects and/or to waive certain laws that protect public lands (such as for the construction of the border wall, as discussed in the conclusion).[32] And allotment policies allowed the O'odham to lease land to outside mining, ranching, oil, telescope, and border infrastructural interests.[33] Although diminished today to one-tenth of its original size, the Tohono O'odham reservation spans 2.7 million acres; it is currently the second-largest reservation in the United States.[34]

At different times, both the United States and Mexico passed laws that offered US citizenship to Native Americans, offers that proved disingenuous at best and more often coercively violent through a promise of uplift engineered to enforce assimilation. After the passage of the Dawes Act of 1887, the US government surveyed Native tribal land to divide it up into individual allotments. Those who accepted an allotment of titled land would be enrolled as US citizens. While the Dawes Act promised to protect Native land from settler theft, the land was turned into private property, severing people from communal relations with land and each other. This law aimed to assimilate Indigenous peoples by channeling tribal membership into US citizenship and responsibilities, veering away from collective and land-based sovereignty toward laws that protected individual families and US national sovereignty. As Patrick Wolfe argues, assimilation is the most efficient settler colonial technique of elimination: "In neutralizing a seat of consciousness, it eliminates a competing sovereignty."[35] After allotments were distributed, any "excess" land was sold to non-Natives.[36]

This trend toward coupling citizenship, assimilation, and land loss along the shifting border line continued when the Indian Citizenship Act was passed in 1924. Citizenship was no longer tied to land title but was offered to (and forced on) all "foreign-born" Native peoples on US territory. Yet US citizenship benefits were unevenly distributed. In Arizona, for example, even those who served in the military were not given the right to vote. And both the US and Mexican governments contributed to diminish O'odham land and rights. After Mexico forced citizenship status onto the O'odham in 1921, only those with Mexican citizenship could occupy land, to keep the O'odham from taking up too much land on the Mexican side of the line.[37] At the same time, O'odham residing south of the supposed border continued to lose land as Mexican officials created agricultural colonies on O'odham land for recent deportees during Depression-era repatriation campaigns. Then the Mexican government turned O'odham land into *ejidos* (land held in tenancy) in 1928, treating them as peasants and farmers who had usufruct rights to land owned by the Mexican state. Now under Mexican rule, they were not considered a sovereign nation like their relatives to the north.[38] Living on ejidos reinforced precarious living conditions as they did not actually own the land. In fact, those who left the land for two or more years, even to participate in the US military during World Wars I and II, returned home to find their land taken away.[39] More recent neoliberal policies during the late 1980s and early 1990s, such as the North American Free Trade Agreement (NAFTA) cheapened the price of agricultural goods such as corn, displacing many Indigenous and mestizo farmers. This widespread economic displacement from their ejidos led to mass migration from Latin America to the United States, including by Indigenous peoples from Oaxaca, where the Zapatista uprising began in 1994, sending Indigenous peoples dispossessed of land in one place onto the land of other Native peoples.

Even today, some see former President Donald Trump's construction of 452 miles of border wall as another settler colonial tactic by the federal government to take land on both sides of the border. This concern is not unfounded as land on the Mexican side of the O'odham reservation has been leased (indefinitely) for the construction of the border fence, and much land on the Arizona side has been taken over by DHS buildings, a detention center, roads, and surveillance towers. In 2017 *USA Today* journalists exposed land grabs at the Texas border.[40] After the passage of the 2006 Secure Fence Act, hundreds of Texans who own property along the Rio Grande's shifting boundaries (many of these families received land grants from the Spanish crown between 1716 and 1836—land that was stolen from Indigenous people) have also experienced the state's takeover of many acres of land while providing little compensation.[41] And since the mapped coordinates of the Texas-Mexico border lie in the middle of the Rio Grande, any border wall would have to be built hundreds of feet north of the line. This means that some will find themselves living on the other side of the border, while others will lose their land to the government through eminent domain.

According to *USA Today*, an estimated five thousand parcels of land, most privately owned, would have to be seized. And this has happened in Texas before. To build almost seven hundred miles of border fencing and walls in 2007, "U.S. officials filed more than 320 federal court actions to condemn private properties. Some cases were settled for as little as $100 for an easement. Others resulted in federal payments as high as $5 million for 6 acres." This theft of land often goes unnoticed because it happens slowly. In fact, "nine years after the first cases were filed with a federal court in Brownsville, 85 remain in litigation."[42]

President Trump has a history of supporting the use of eminent domain during the 1990s in New York City under the auspices of city development. What he really wanted was to evacuate people to clear land for his limousine parking lot at the Trump Casino. The use of eminent domain came from a law passed during the Depression—the Declaration of Taking Act (1930)—aimed to stimulate the economy. This law allowed the federal government to seize land quickly. This land grab mostly displaced Native, Black, and poor people. Most famous were the inner-city evictions in slums or blighted regions during the 1950s and 1960s, called *urban removal*, which James Baldwin renamed "Negro removal."[43] Eminent domain is also used to establish military bases, mining, "public-use" projects such as national parks and observatories, prisons and detention centers, and gas pipelines. For example, since 1917 O'odham land has been occupied by US military and government projects such as the Organ Pipe Cactus National Monument, the Luke Airforce Base and Goldwater Bombing Range, a copper mine in Ajo, the Kitt Peak National Observatory, and the US-Mexico border.[44] In a savvy reversal that may reverberate to other reservations, on October 23, 2016, the Standing Rock Sioux reservation declared eminent domain on their land that was under threat of being occupied by the Dakota Access Pipeline project.[45] As we will see in the conclusion, national security is deployed as a "public good" that confounds the boundaries among care, destruction, conservation, extraction, and security to usurp more O'odham land.

Black Holes and Security Voids on the Border

Just as the state was complicit in awarding settlers land considered empty or unoccupied, the DHS mandate now is to occupy and control border zones that are *devoid of security*. During the 1970s, media articles reflected the growing fear that the Tohono O'odham reservation was a "borderless" border where illegal aliens could easily cross given the lack of security personnel, checkpoints, or even border signs warning would-be crossers that they were now entering US territory.[46] To the O'odham, this was not, and is not, US land. Yet, with increased immigration, cartels increasingly encroached onto O'odham land in Mexico, causing many to flee north.

The influx of drug runners onto the reservation led the O'odham tribal council, against the wishes of many O'odham tribal members, to work with Immigration and Customs Enforcement to develop an all-Native Border Patrol called the Shadow Wolves in 1972. Two years later, the O'odham council agreed to house a Border Patrol office on their reservation and then a vehicle barrier in 2007–2008. Despite these compromises, the association of the reservation border with a security vacuum continues. Newspapers repeatedly characterize the O'odham reservation as being the size of Connecticut while having only thirty-four thousand members. The "innocent" repetition of this "objective fact" in just about every newspaper article about the O'odham inadvertently enacts discursive violence, the violence of rendering Native land as an empty or unpopulated region, a land in need of state control and occupation. As a desert terrain, supposedly beyond the reach US sovereignty, this land is once again characterized as inhospitable, a death trap for immigrants, and humans more broadly, who attempt to cross on foot. In fact, an *ABC News* account reinforced the idea of a security void along the O'odham border, describing it as the "widest open space to patrol" and a "hot spot" for "Mexican drug cartels and human smuggling."[47] As stated by a US Border Patrol agent, "The Tohono O'odham is one of our most problematic areas. The narcotics smugglers have moved up into the mountainous area. There is not a lot of access." Similar to the Apache from chapter 1, who easily navigated the mountains to escape the cavalry's reach, here too, border security is necessary to prevent smugglers from hiding on O'odham land.[48]

In other media accounts, similar border regions are labeled "black holes," or "an enclave largely beyond the control of authorities on either side of the border because of its remote location."[49] According to the state, these regions are lawless spaces that attract and proliferate illegal activity by the most reviled populations, such as drug runners, gangs, immigrants, and cartels; in these spaces, referred to as "Indian Country," the laws of nature are deformed, or made perverse, allowing dangerous and foreign elements to be propelled into the nation, sucking more deviant populations into its ever-increasing folds, until the takeover is complete and all life slips into fecund darkness. This colonial fear of the dark and wild regions at the edge of state control and Western knowledge meets future fears of a "silent invasion" if illegal actors are allowed to swarm freely. Again, for the O'odham, it is not the illegality vacuum that threatens their extinction, but the imposition of US state sovereignty and the intrusion by the DHS and other police-military-border security forces.

Journalists and popular media outlets contribute to a cult desire for militarized borders, merging reality television with embedded journalistic accounts that naturalize the militarization of everyday life on the border. These shows also validate a relentless expansion of funding for border-security technologies, personnel, and infrastructure.[50] These accounts reinforce the scientific-secular belief that seeing is believing, that there is a truth to the chaos of the border, that it is visually apparent, and that militarized security is the only way to stop its spread.

USA Today recently launched the Pulitzer Prize–winning project "The Wall," where viewers interact with the border virtually on foot or by air, in an attempt to expose every crevice of the border. Not only can you explore the border through virtual-reality goggles, but they also use GPS, helicopter footage, and digital and thermal mapping technologies to visually document the entire border, including where fencing and walls begin and end. Their goal? To provide viewers a complex understanding of what it would mean to build President Trump's wall along the entire 1,954-mile border. In contradistinction to early surveyors, who were technologically and humanly unable to map the border along the rough Rio Grande or in much detail, given the limits of passage across mountainous regions or across Indian territories, the *USA Today* team of journalists exploited the latest technologies to produce one of the most extensive border maps ever made.

One of the many video reports and podcasts available on the website for "The Wall" is "A Border Tribe and the Wall That Will Divide it."[51] Set on the O'odham reservation, the video begins with the visual image of a border fence and line marked in the sand stretching far off to the right of Verlon M. Jose, the vice chairman of the Tohono O'odham Nation. Despite the presence of a border fence visually dividing the land, the video begins with Jose singing a song of prayer in their Uto-Aztecan language that travels far beyond the border fence, crowding out its logic. Songs of prayer breathe sacred messages into the wind, or the life of the universe, merging with ancestral voices that have lingered and only now find form to take flight on swirls of wind that carry the past into the present and future, awakening the land to the presence and care of the O'odham peoples. There is a constant exchange of presence that echoes back, such as when, during our visit, the sound of the concha and our prayers sent up to Baboquivari Mountain moments later refracted back at us from the mountain in a slightly altered soundscape of communing.

Chairman Jose says, "In our way of life, we were never put here to be in one place. We travel. From place to place. Wherever the water is, wherever there's food growing, wherever the animals go. We didn't cross the border, the border crossed us." As the camera spans a reservation that suffers high unemployment and poverty, we are led to question the camera's visual archive documenting an impoverished, barren land. The spirit of the O'odham people overwhelms the camera's harsh colonial eyes. At the heart of the O'odham way of life is movement. In contrast to the perception of the border as a stable line, the land is in perpetual motion, modifying itself depending on weather patterns, animal migration, and human behavior or intervention. As a living force, it requires rest, honored by the seasonal migrations of humans and animals. Jose continues, "We must keep this traditional passing, regardless of what's before us. The creation of a wall, the creation of a barrier, will impede the flow of life. One thing it will not stop is our prayer. The O'odham say there is no language for a wall, so you're creating something that does not exist." When Jose says the creation of a barrier will impede the flow of life (Him'dag), he speaks to the heart of sovereignty in this region. The movement of O'odham people across the artificial border is tied to their social, cultural, and ceremonial or religious practices. These pilgrimages follow the path of their ancestors to imprint memory and to tap into sources of power that can powerfully heal past and present wounds against the people and earth and help them to thrive economically, socially, and spiritually. As stated by Rafael Antonio Monreal later in the video, their pilgrimage to the salt mines in Sonora carries the knowledge of the ancestors: "The salt mines, they have been here for a long time. It was used for different purposes, for medicine, for preserving our food to carry over from village to village to different parts of Arizona and Sonora. Our ancestors walked this land for eighteen to nineteen hours a day. And we tried to follow up the trails that our people did. Same mountains. Same cactus. Same sun. Same heat. Just one wall, it will make a difference for all the O'odham, for all of mother earth."

On the website *O'odham Solidarity Project* is a manifesto, "O'odham VOICE against the WALL," that lists the ways the DHS and Border Patrol have violated the O'odham (1) "Right to Life," (2) "Cultural Rights," and (3) "Right of Mobility" as well as some of the ways the DHS presence (4) "Trespasses on and Destroyed O'odham Cultural Property."[52] This list intentionally refers to the UN Declaration on the Rights of Indigenous Peoples adopted in 2007.[53] The DHS violates O'odham rights to life by continuously monitoring and surveilling communities on the entirety of O'odham lands, restricting free movement within communities and entire lands, engaging in armed abuse of and violent attacks on O'odham members, and driving into yards and fenced-in areas at high speeds, endangering the lives of O'odham.[54]

The reservation has become a police state where O'odham are constantly under scrutiny.[55] Those who speak out against the DHS or the Border Patrol have even experienced dire effects, from being treated violently to having their Social Security check payments stopped. The process of becoming bordered by the militarization of their land involves a disorienting experience of having to declare US citizenship, rather than tribal membership, and of constantly being seen as suspicious, as a potential migrant, drug trafficker, criminal, or terrorist. All the everyday practices (language, food, hunting, ceremony, praying, even a spirit of generosity) that shape who you are become something else entirely.

This disorientation extends beyond the O'odham and expresses the disharmony across space and time for all living creatures. Jose comments on the disorder that a partial wall and security regime have caused in their natural surroundings. He says, "The rattlesnakes don't know what season it is, the saguaros aren't blooming on schedule. . . . It is what humans are doing to Mother Earth. In order for the world to be in balance, Mother Nature has to be in balance, too."[56] Careful observation of animals, plants, and the land is part of the skills necessary to read the patterns that communicate the earth's temporal disjuncture, imbalance, and ill health. To be oriented through O'odham Him'dag is not to imagine oneself in relation to mapped grids, or along a progressive historical timeline. Instead, to be oriented is to place oneself within a sacred order that expands from below the earth to the cosmos, grounding us in a time that is in sync with the path of the ancestors and in alignment with the four directions. Currently, this sacred positioning and vision are challenged by a secular-military optic that uses its targeted gaze to alienate or to strip one of a sense of self that ties people to forces all around them, including the harvest of saguaro fruit, which we will hear more about in the conclusion. The nation-state and borders here work together to cleave self from other, inside from outside, good from bad, and so on. The gathering and harmonizing work of cosmological notions of self-in-relation privileges an attunement to life that stitches together selfhood with the health of one's surroundings.

During the second half of the twentieth century, an increase in border control disallowed O'odham from the north from migrating south during periods of increasing drought, thus barring them from accessing better water and agricultural land, as well as shared resources with family on the Mexican side. Thus, the border crisis and climate change crisis become coproduced. That is, border occupation exacerbates climate change by dismantling O'odham tribal members' ability to move, thus stripping them of their land-based livelihood. Even though they are not physically removed from their land, they are invariably forced to assimilate by becoming laborers in the cash economy. Yet even the labor economy has proved precarious. For reasons as diverse as cotton mechanization and the increased migration of Mexican, Black, and Central American migrants, O'odham were pushed out of the labor economy from the 1940s to the 1960s. During this period, the O'odham population surged on the US side and decreased in Mexico, while the O'odham family income dropped to one-fifth that of whites.[57] The 1990s witnessed an increase in jobs in Arizona, opening up more opportunities for northern migration, yet also further depopulated southern-based O'odham, reinforcing unequal access to labor markets for O'odham on both sides of the border.

The fortification of the border today prevents the O'odham on the Mexican side of the border, as a people unrecognized by the Mexican government as either being sovereign or having citizenship, from crossing into the north to access important resources—such as hospitals and clinics—that O'odham on the US side enjoy. Since the O'odham decide membership based on ancestry (and not country of origin), the southern O'odham have the same rights as those who happen to live north of the border. And while only two thousand or so live on the Mexican side, this number is much smaller than in the past, owing to cartel and drug violence in Mexico, which have caused many to abandon their homes and flee to the US side of the border.

The loss of movement is akin to the death of a people and their ritual practices that sustain an entire way of life. Some say the threat of a wall will arrest practices that hold memory and transmission and thus lead to their people's extinction. While the term *extinction* usually refers to the disappearance of plants or animals, here it refers to the multiple webs of interdependent life that will perish once a border wall is constructed and security operations perpetuate occupation. Animals will be prevented from following their migratory paths, water will flood rather than flow, connections to ancestral knowledge will wither, and the movement of the O'odham on both sides of the border will be arrested or slowed. In comparison to the term *genocide*, which refers to the extermination of a human population, *extinction* highlights the complex webs stringing together humans with the land and natural world so that when one species goes extinct, this diminishes entire pathways of life for all. In this vein, Audra Mitchell argues that extinction is not like genocide since it is not intentional and thus cannot be directed at a particular population. Extinction, Mitchell argues, is not a metaphor; it *is* genocide. Extinction = genocide thus enacts particular ecological patterns of structural violence that disproportionately affect specific racialized groups.[58] She gives the example of the time when the buffalo were almost eradicated and, along with the buffalo, an entire First Nations people whose way of life was tied to the migratory paths of the buffalo.[59] On the O'odham reservation, the presence of the DHS continues colonial patterns of settler violence that conspire to cut off the flow of life, stripping away the possibility of Native sovereignty, mobility, and land-based practices and thus access to an entire way of life.[60]

The Militarization of the Sky over Sacred Land

When O'odham members demand that the DHS de-occupy their land, they detail the ways in which border-security presence, surveillance technologies, and even foreign languages assert a material violence and worldview that impede their sovereignty and existence. Just a few days before I arrived at the O'odham reservation, the tribal council negotiated with the DHS to erect nine of the sixteen proposed integrated fixed towers manufactured by Elbit, the Israeli corporation that developed drones to monitor Palestinians.[61] This "collective" decision went against what many O'odham tribal members wanted, knowing this resolution included a promise by the DHS to pay the reservation money to lease the land for twenty-five years, while about seventy miles of roads would be built to erect the towers.[62] In an online site, O'odham elder Rivas decries the building of towers on their sacred mountains. The construction of zigzagging roads up the mountain will damage ceremonial mountains, lands, and hillsides and destroy the natural habitat that is of great use for hunting or harvesting medicinal plants, while also desecrating burial sites.[63]

These hovering towers will block the community's view of the sacred mountain and obstruct the vibratory direction of prayer songs and breath carried by the wind that transport their words and thoughts up toward the mountain, the sky, and beyond. These prayers are sacred gestural codes that regenerate ancestral practices and require a clear path for the mountains, sun, clouds, and rain to hear or sense the cues—human and otherwise—that trigger their mutual communication. The towers have other negative effects on the animals and people, such as severely disturbing the delicate sensory fields of the bees who produce their food, for example. Even though in March 2017 the US Customs and Border Protection issued an environmental assessment and found no significant impact of the towers on the Tohono O'odham Nation, O'odham elder Rivas conducted a study with two other researchers on the consequences of the towers.[64] They state that the agency found no harm because it "does not understand environment as holistic, interconnected, and living. It also does not fully disclose facts about the towers' impact."[65]

From data collected by Caitlin Blanchfield and Nina Valerie Kolowratnik, the 120- to 180-foot-high surveillance tower currently standing in Chukut Kuk District on the Tohono O'odham reservation can track moving people within a 9.3-mile radius and moving vehicles within an 18.6-mile radius, as well as providing long-range video within a 13.5-mile radius.[66] Their map makes visible the hidden reach of radiation by showing circular rings emanating out from only the first of nine proposed tower locations.[67] The surveillance cameras are so powerful they can see through walls and into cars. Other border surveillance, such as the border AVATAR scanners described in chapter 3, scan border crossers' bodies to collect data on each person's physiology. In both cases, the state penetrates into the private recesses of the reservation and tribal members' homes. This technology may one day be used to collect data on their bodies.

Not only is their land dissected by roads, checkpoints, and border-security personnel and vehicles, but the sky is littered with helicopters, drones, and airplanes during the day and at night. This has interfered with and disrupted ceremonies and hunting practices as helicopters fly over these areas, scaring prey and spotlighting ceremonial dancers. Even local hunters, who rely on the food they catch, are harassed and treated like criminals for wielding guns and knives when they are caught by remote sensors and border officials driving through the reservation.

My short time on the reservation gave me a sense of the unnerving presence of Border Patrol vehicles and the constant assault of Air Force planes and Border Patrol helicopters and drones. It's difficult to understand the visceral and psychic invasion of this militarized presence if you haven't been on the land and if you have no idea what makes the O'odham such a special people who have tended and made sacred their land. As an outsider, I am limited in my understanding, although I did sense some of the sacred design of this land. I could hear footsteps crunch in the sand from far off, a sound that is muted by concrete sidewalks and asphalt roads in cities. Sounds reverberate more forcefully because the reservation is surrounded by mountains. It was quite an experience to be on land where I had a clear 360-degree view of the sacred mountains that hold the O'odham together through their stories, protection, and gifts of water and nutrients.

When we arrived, the spring desert was in full bloom, a riot of colors stunted only by asphalt roads and otherwise spreading across the land as far as the eye could see. Flowers returned that had gone dormant for over a decade, and plump animals such as donkeys, cattle, and horses roamed across the land with few fences besides those at the border. Naming this land a desert fails to hold the abundance and sacred symmetry of land many O'odham consider holy.

This delicate soundscape exaggerates the sonic violence of the militarized aircraft tearing across the sky (much too close to the land) in what the US government considers sovereign airspace. I was literally jolted by the sounds of helicopters blasting away the quiet of the desert. It's not the same in cities, where there is a sense of wonder at the sight of a quiet plane flying far above. Even while on a run one morning with a few O'odham youth, both Iriany and I shared the sensation that we heard large flapping wings behind us. We were amazed that both of us turned to look behind to see what could be making this sound. Despite not seeing anything, I felt my footsteps light, as if aided by a large bird lifting me slightly. It was as if my momentum was also being pulled forward when running toward the sacred Baboquivari Mountain. And during our runs I could feel the daunting effect of large aircraft above leaving a dark shadow across the otherwise clear view of the mountains. These aircraft are a constant reminder of not only US technological occupation of the sky but the ability of helicopters and drones to see people on the ground from a vista high above. Some drones are equipped with electro-optical sensors (cameras) that can identify an object the size of a milk carton from an altitude of sixty thousand feet.[68] This is an unnerving reminder that one's every move is in intimate proximity with the state, a presence that threatens to block sacred sensing with harsh accusatory eyes. By making the land and bodies into small objects viewed from above, surveillance attempts to capture life at an alienated scale that easily translates phenomena into knowable data bites, into information or images. The directionality of knowing is reversed, as sacred vision respects all scales, even life at the most minute scale, as being in motion, in dialogue with the sky and underworld. A view of life from above renders impossible what can't be seen or known by this technologically reductive visioning.

It may also be helpful to think of the state's occupation through what Arturo Escobar calls *ontological occupation*, or the threat that one's very collective being, an intrabeing with all life, will be alienated or broken apart. And this relational ontology entails the theft of a people and the land. A key aspect of Escobar's understanding of ontological occupation is autonomy, which he borrows from Francisco Varela, who says, "The key to autonomy is that a living system finds its way into the next moment by acting appropriately out of its own resources."[69] If the O'odham cannot commune with the mountains, nor protect this being from violent intrusions, the material and ephemeral flourishing of life on land and in the sky is blocked.

The surveillance towers are also a constant irritant in the soundscapes of residents, who are reminded of the buzzing surveillance and security presence above them. How can the Tohono O'odham Nation protect its sovereignty when cartels use airplanes to airdrop drugs onto the reservation or when Immigration and Customs Enforcement drones fly not just along the border but over their entire reservation? What kind of protection is there for safeguarding not simply land access but their sacred airways from surveillance drones, which never touch the reservation? As argued by Derek Gregory, "The capacity of drones to conduct 'air policing' reactivates a colonial form of power in a radically new constellation. For the drone makes possible an extended *occupation* rather than a time-limited *incursion*."[70] And "the ingenuity of this novel form of aerial occupation," Campbell Munro suggests, is "its capacity simultaneously to respect and transgress the principle of territorial sovereignty."[71]

As is happening across other reservations, they struggle to protect their sovereignty from the intrusion of drone operators such as the non–tribal member Lionel de Antoni, who in January 2009 flew his drone over the reservation of the federally recognized Hualapai Tribe in the Grand Canyon without the tribe's permission and posted pictures on his website to sell tours.[72] Given the Hualapia's reliance on tourism, the tribal police confiscated his equipment. De Antoni argued that he never physically touched Hualapai land since he launched from and landed onto federal land. His lawyers stated that the federal government has sole jurisdiction over the nation's airspace. While no legal battle ensued, as he opted to pay the tribe a fine, this case raises deeper questions about whether tribes, as sovereign nations, can control their own airspace. Across Indigenous land, locals put up signs demanding that tourists not fly drones over their homes, burial grounds, or other sacred land. A similar legal battle over aerial sovereignty is ongoing with Sioux members who call themselves "Digital Smoke Signals." Declaring themselves "eyes in the sky," they flew drones over their own land to document the violence and illegal acts of the Dakota Access Pipeline construction at Standing Rock.[73] While footage from a private security helicopter was used by police to surveil and convict protestors, once the Federal Aviation Administration realized that members of the Standing Rock Sioux tribe were launching their own drones to gather video footage of law enforcement, they temporarily halted the use of drones to create a media blackout. This is similar to the 2014 temporary flight restriction put in place in Ferguson, Missouri, during protests over the police killing of eighteen-year-old Michael Brown, a move intended to keep the media out.[74]

The Right to Mobility

Border security on the reservation has also meant that in order to practice their way of life, O'odham must carry a legitimate ID with them at all times. This proves difficult when some O'odham, like my own grandparents in Mexico, lack birth certificates. This means that even the original inhabitants of the land become undocumented, similar to immigrants. Yet, like the use of the monarch butterfly in migrant activism, it is impossible to stop the flow of movement by all life's forces: "When the wind blows, they're gonna stop it and ask it for papers? When the water flows, they're gonna stop it and ask it for papers? And all the animals that have migrated for thousands of years across the border?"[75] The right to mobility is simply a part of life, a flow that cannot be stopped by the state. Still, many are violently restricted from moving, their land a carceral space, just as the Palestinians' land is occupied by Israeli checkpoints and surveillance towers and surrounded by massive walls. In fact, upon leaving the reservation, O'odham have to declare their citizenship status.[76] The official O'odham website confirms that O'odham members have been detained and deported for not having the correct identification while crossing between Mexico and the United States on their own land.[77] And given that O'odham on the Mexican side are not recognized as US citizens, they are also trapped on their land, harassed by the Mexican military, and authorized to enter the US side of the reservation only to use the medical facilities. Many are prevented from visiting their families on the other side, participating in ceremonies or fiestas, and even sharing traditional knowledges with others.

Sovereignty has been compromised not only at the reservation's border with Mexico but also across the entire reservation. Simply being O'odham is criminalized. Many describe being apprehended by the Border Patrol, and even having a gun put to their head, after being mistaken for an illegal migrant. For this reason, some local residents hide their fluency in Spanish on the reservation for fear that they will either be mistaken for migrants or be regarded (even by other locals) as someone who abets migrants.[78]

Despite ongoing protests, helping migrants—by offering water, food, lodging, or a ride—has become a criminal act. The uneven consequence of ignoring US laws is not missed by O'odham members, who know that when white citizens offer migrants water, it translates as "humanitarian assistance."[79] Although many feel strongly about helping migrants on the reservation, the tribal council finds itself in a difficult position when it has to ban residents from practicing their core values of helping others, in this case by leaving water in the desert.[80] Council members say they have to negotiate with the

US government to protect the resources and support they receive to handle the onslaught of drug runners and migrants. The tribe spends about $3 million annually on border issues, funding that could be used for their own people. Most expenses related to migrants include emergency care, autopsies for those who die on the reservation, towing of abandoned vehicles, and the construction of holding cells or detention centers. In one of the most comprehensive reports on the border in 2003 (over a hundred pages long), the chairman of the O'odham Nation and a local chief of police addressed the problems faced by the nation when local resources go toward the border. The reservation ends up, for example, lacking the resources to handle the increasing numbers of young people addicted to the drugs trafficked across their reservation. Youth and their families end up with costly bills after checking out of Arizona hospitals.[81]

From Indian Scouts to Shadow Wolves

While the ongoing militarization of the Tohono O'odham land has a long history with the United States that began in the mid-1850s and has ratcheted up since the 1970s, we can see the resonance with the Apache Wars, discussed in chapter 1. In 1972 Congress agreed to the O'odham Nation's demand that they would consent to having Border Patrol on their land only if the officers were from the O'odham reservation. The seven original Native American trackers, who called themselves the Shadow Wolves, were hired from the O'odham tribal police to help prevent drug smugglers from moving drugs across the reservation's border. While this original group all lived on the O'odham reservation, after 9/11 another group changed their name to NATIVE and expanded their membership by hiring members from other tribes, such as the Blackfeet, Lakota, Navajo, Mohawk, Omaha, Sioux, Yaqui, Pima, Kiowa, and Yurok.[82] One of the first hired in 1972 was Stanley Liston, an expert scout who could "walk into a campsite where smugglers were sleeping, count their weapons and walk out without being noticed."[83]

Similar to the Apache from chapter 1 who could move through rugged terrain without detection, a more recent recruit, David Scott, is an Oglala Sioux from South Dakota whose "great-great grandfather was a U.S. Cavalry Scout and one of the first tribal police officers to patrol Pine Ridge in the early 1900s."[84] This long legacy of Native skills as hunters, surveyors, and warriors could be interpreted, on the one hand, as a key asset in the arsenal of militarized border control and war. On the other hand, Native trackers use the skills passed down from their grandparents to protect not US territory but their own

sacred land from drug runners who they say trample through the ancient cem-
eteries and holy places:

> That peak up there, he says, speeding toward the reservation's border
> with Mexico. That's where the creator lives. His name is I'itoi, the elder
> brother. He created the tribe out of wet clay after a summer rain. Tribe
> members still bring him offerings—shell bracelets, beargrass baskets and
> family photos—and leave them in his cave scooped out of the peak. But
> the drug smugglers don't know that. On their way to supply America's
> drug markets, they use these sacred hilltops as lookouts, water holes as
> toilets and the desert as a trash can. . . . I like to think I am protecting
> not only the U.S. but my area as well, my home.[85]

Rather than put the blame on border security, drug runners and even migrants are blamed for disrespecting their reservation home. During another
run on the reservation with a couple of O'odham youth, they hardly needed to
point out the many signs of migrant crossings—from towers with emergency
buttons for migrants to push in case they wanted to be rescued by the Border
Patrol to rug-soled slippers (used to cover their footprints) strewn across the
desert floor. The land held all these traumatic objects and their painful stories,
an open wound leaving traces of trauma for the O'odham to navigate.

Smithsonian journalist Mark Wheeler followed Nez, a Navajo Border Patrol
officer working on the O'odham reservation, as he tracked a group of smugglers
carrying bundles of drugs onto the reservation. Touted as using ancient tracking skills, knowledge passed down from their grandparents, or even knowledge
from current hunting skills, or from tracking escaped livestock, Native trackers take pride in (and are acknowledged in popular culture as) having skills equal
or even superior to the expensive high-tech surveillance technologies used by
US Border Patrol. Their ability to track and hunt animals and humans "like a
small pack of wolves" translates Native knowledge practices into useful skills
on the border. Rather than relying solely on gadgets such as night goggles or
remote sensors, they also "cut for sign," or search for physical evidence in the
terrain—"footprints, a dangling thread, a broken twig, a discarded piece of
clothing, or tire tracks."[86] For such a small group of anywhere from several to
twenty-one members, they have had great success in confiscating drugs and
finding the smugglers, who themselves use mountaintops, high-tech communication technologies, and countertechniques to conceal their tracks. The
spread of these techniques to the US Border Patrol is countered by numerous
inventions by drug runners and migrants to evade detection, such as covering

their feet with pieces of rug, placing rubber stamps with a cow hoof on the bottom of their shoes, or even blowing their footprints away with leaf blowers.[87]

Much impressed with their techniques, Wheeler says the shadow wolves "cut sign like other people read paperbacks."[88] The reference here to reading the natural terrain like a book to glean knowledge situates the land as a stable sign from which to abstract knowledge, and Native knowledge as prelinguistic, hovering on the border to the primitive and yet superior to modern technology. Popular media on Native tracking refer to their ability to read the land like an ancient hieroglyph, or inscrutable image whose meaning eludes Westerners. Wheeler's *Smithsonian* article details the journalist's experience with the Shadow Wolves who see marks on the terrain—hieroglyphics that are imperceptible to him, including "nearly indeterminate scratches in the sand." For Nez, these hieroglyphics are clues to be read along with knowledge of the celestial movements of the sun as it marks shadows and light on the terrain. In one case, he interprets marks in the dirt as key evidence of someone resting with a bale of dope. Nez translates his knowledge of the sun's position to guess the distance traveled since the smugglers moved when the sun forced them from their shady resting spot. When I was on the reservation in March 2019, I noticed that my own footprints were easily detectable in the sandy ground when the sun was angled in the right position.

The almost mythical ability of the Shadow Wolves to move undetected, and to see what cannot be seen by most human eyes, has generated abundant material for films, novels, and extensive journalistic accounts. As Western vision of land is increasingly abstracted through televisual, filmographic, and drone footage from above, this intimate relation with the land becomes another dark colonial mystery, a black hole in the Western imaginary. In fact, Native Border Patrol officers are so effective that they have been hired to search for Osama bin Laden and Taliban terrorists in Afghanistan and to teach other border personnel the science of "cutting sign." These skills have also been deployed to catch drug runners in border regions of Latvia, Lithuania, Moldova, Estonia, Kazakhstan, Tajikistan, and Uzbekistan (which border Afghanistan).[89] The DHS converts tracking into a one-dimensional "scientific" knowledge that abstracts this knowledge from its spirited connection to place and transforms it into a global technique of warfare across all borders.

Immigration and Customs Enforcement is in the process of forming another Native Shadow Wolves unit at the Blackfeet reservation in Montana, which borders Canada, to address increasing cross-border drug trafficking.[90] And the Mohawk reservation that borders Canada and New York is also suffering from the increased militarization of the Canadian border, including the undermining of the sovereignty of the Mohawk Nation by US and Canadian Border Patrol, New York police, the Federal Bureau of Investigation, and those who are targeted by the nation as "terrorists." In fact, the Mohawk launched a protest demanding that the United Nations send officials to Akwesasne to de-escalate the use of guns by the Canadian border patrol and the increasing harassment of Mohawk who cross the US-Canada border as part of their normal practices, which are supposedly protected by treaty agreements with Canada. The protest recognized that this border was to remain "undefended" according to the Jay Treaty of 1784 between Britain and the United States.[91] Similar to the case of the Tohono O'odham Nation, the high poverty and unemployment rates on the Mohawk reservation make it easy for the youth to be seduced into the drug trade.[92] And the increased presence of drugs on the reservation has led to increasing numbers of youth hooked on highly addictive painkillers such as OxyContin. This then leads to petty crime to keep up with the expensive doses.[93]

In fact, the deluge of drugs brought onto the reservation causes severe problems for the O'odham youth, some of whom become hooked on drugs. And border security continues the slow removal of young people from their families and their land. In an interview published online, elder Rivas said that when young people are sent off the reservation to serve time in prison or detention, many are forced into halfway houses instead of coming home.[94] "When family members do not come home, families wait, as there is no system set up to find information from the Department of Homeland Security."[95] All of these issues are complicated by the lack of legal jurisdiction and thus adjudication of non-Natives who commit crimes on the reservation, and the tribal nation's inability— whether through the US criminal system or health care—to handle the consequences of drug cases affecting almost all O'odham families on the reservation.[96]

Conclusion

Given the long history of Indigenous dispossession on the frontier between Mexico and the United States, the declaration of a security void and border "crisis" on the O'odham land is another tactic of colonial occupation on O'odham land and an affront to their sovereignty. At the same time, O'odham members mitigate visual containment by engaging in what anthropologist Reyna Ramirez discusses as "spiritual unmapping."[97] By continuing their spiritual practices and relation with their homeland, they assert their own forms of belonging with land in ways that surpass the limitations imposed by technological devices that can be interrupted, that fall apart, that rely on constant upgrading and resources, and that can be washed away during the flood season.

As Jose stated, "We carry the spirit of thousands of warriors. Dominant societies have tried to kill the spirit of our people. . . . We will survive."[98]

In this chapter I have complicated the meanings of mobility, migration, and
citizenship by assessing how the settler border-security regime and its surveillance technologies create more precarity for migrants but also for the O'odham
tribal members whose land they cross. Not only are migrants detained and
deported by the Border Patrol, and oftentimes killed by the funneling of their
movement into the desert, but also the O'odham who live on the land they
cross en route to the United States. It is critical, then, to take into account
the wider ecology of migrant movement and criminalized containment by
the settler state when advocating for migrant rights and access to citizenship.
For many Indigenous people, the promise of citizenship has been yet another
tactic of settler occupation and dispossession that naturalizes state claims to
sovereignty over that of the O'odham. Among the O'odham who manage to
live outside the drug trade, and under the state's radar, many refuse the militarization against migrants, just as they resist their own militarization. This
onslaught of security threatens to sever the webbed rhythms of all life, just
as the displacement of migrants sets them on a precarious path to the United
States. Just how this security infrastructure, research, and development took
hold in Arizona is what I turn to next.

THREE

Automated Border Control *Criminalizing the "Hidden Intent"*
of Migrant/Native Embodiment

In the previous chapter, I traced the conflicts faced by the Tohono O'odham nation under conditions of occupation by the Department of Homeland Security (DHS), including the Border Patrol, ground sensors, aerial surveillance towers, and helicopters. The O'odham, similar to the Palestinians, are under the duress of an imperial power that demands full access to their bodies and land. This chapter turns to another settler colonial border technology, AVATAR (Automated Virtual Agent for Truth Assessment in Real-Time), an automated Border Patrol agent armed with sensors developed at the University of Arizona, tested on the Nogales border, and exported to borders around the world beginning in August 2018. It extracts physiological data from border crossers in the hope of arresting the movement of anyone showing signs of suspicious gait hidden in the body's unconscious movements.

The AVATAR kiosk, armed with over fifty biometric devices, rapidly scans a person's body with sensors that measure physiological movements and behaviors (such as eye movement and gait) in search of the "truth" of one's identity and one's possible future actions, further stripping border crossers of the human right to movement (and residence) along the border (see figure 3.1).[1] By segregating the body from collective stories that tie one to ancestral presence on the land, surveillance once again not only monitors migrants but enacts settler control over the borderland through access to each person's embodied data footprint. In the name of security, AVATAR collects and archives data from travelers' bodies at border checkpoints, following one's archival record for an unknown duration of time. Surveillance thus extracts information from bodies as the novel terrain for governance, sovereignty, and profit. Border officials usurp the power to translate this data into binary codes of threat and nonthreat, with dire consequences.

Arizona is a hub for designing visual sensors as the field of optical science attracts businesses from far and wide, leading to the region's popularity as

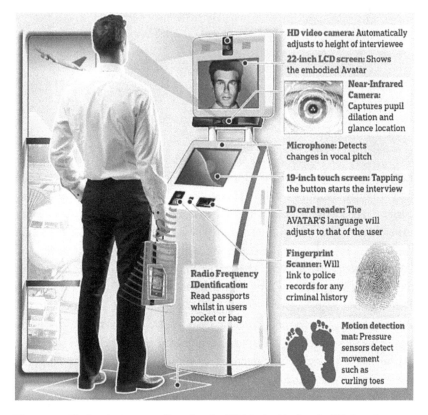

Figure 3.1. The border AVATAR. Tested at the US-Mexico border, the US-Canada border, and various European Union country borders. Source: *Kozminski Techblog*, http://techblog.kozminski.edu.pl/author/40963/.

"Optics Valley."[2] Mount Graham International Observatory houses the largest single telescope in the world, located on a sacred mountain referred to as Dzil Nchaa Si An (big seated mountain) by the San Carlos Apache, who continue to fight to protect the mountain from infrastructural occupation. On Baboquivari Mountain (Kitt Peak), sacred to the O'odham, twenty-one telescopes are contested through an ongoing lawsuit by the tribe. These sites, and a sprawling border optic surveillance industry, are part of the research arm of the University of Arizona. The Apache, O'odham, and many other tribes contest Western claims that scientific feats of exploring outer space are more important than protecting their sacred mountains where their stories as a people begin, where ancestral burial grounds are located, and where medicinal water, plants, and spirits reside. These high points are important religious portals connecting the earth with the sky. Arizona vies for dominance in the neocolonial race to see into the unknown, from outer space to the dark recesses of the potentially threatening body, thus secularizing sacred relations among body, land, humans, and the more-than-human. Western science must constantly innovate to eradicate the unknown, positioning Arizona as the beacon of light driving optical technologies that aim to see and know phenomena at all scales. By surrounding the border, mountains, and checkpoints with optical sensors, many imported from Israel, Arizona amasses profit as a global laboratory for the carceral containment of potentially rogue threats snagged by its optical border biosecurity-industrial complex.

As I follow the development of an optical sensor technology targeted to apprehend Mexican migrants, it is imperative to hold on to the settler colonial context of Indian threat programmed into the sensors' vision. By taking us back to the development of automated sensors by university researchers and corporations in Arizona and Israel, I follow the research and development behind AVATAR, including a field of study called *deceit detection* that builds on Sigmund Freud's nineteenth-century theorization of the (primitive) unconscious. I argue that the primitive unconscious driving these automated sensors in Arizona continues to rely on Nativision, or the eyes of the Indian scout who tracks, once again, not the inscrutable signs of presence across the land but the wild or deviant data traces found in the unconscious flesh of the body. For Freud, a lingering attachment to primitive feelings, desires, and thoughts could be drawn out of a patient, and eradicated, through verbal discussion of dreams or other suppressed desires. While Freud required patients to speak about their secret longings and fears, today AVATAR incites and documents what Michel Foucault would call a "regime of truth" from the unconscious movements of the body.[3] Similar to the psychoanalyst, automated sensors reach into the body's seemingly hidden interiority to contain (detain and imprison) and eradicate (deport) uncivilized, or primitive, desires, now rendered knowable through the feats of technoscientific advancements.

I seek to understand the techno-racial unconscious driving automated approaches to border control, which are premised on the idea that the hidden physiological movements of the body are more reliable than speech for detecting whether one is telling the truth or lying, whether one is a potentially good citizen or someone who is deceptive, untrustworthy, or deviant. September 11, 2001, provided a fertile context for the emergence of fear of deceit, a fear that fueled a spike in spending on border security, surveillance, and biometric technologies such as the development of AVATAR.[4] Following the discovery that the terrorists who hijacked the planes had used false documents, the US government unleashed a campaign against identity theft at home and abroad. In fact, in 2001 identity theft was labeled the "fastest growing crime of any kind in our society."[5] Armed with the growing fear of document fraud by terrorists, the government implemented biometrics (fingerprint, DNA, and iris scans) to verify a match between one's bodily imprint and the identification one carries. In other words, by turning to biometrics, it hoped to answer the question, *Are you who you say you are?*[6]

We are entering an era of automated border control shaped by an emerging optical border biosecurity–industrial complex. In a statement on behalf of the DHS, then secretary of homeland security Janet Napolitano described her agency's goal of taking "a risk-based, intelligence-driven approach to help prevent terrorism and other evolving security threats."[7] The metaphor of *evolving* harm across leaky borders speaks to the biological hue that colors threat as an inevitable, and continual, natural disaster that has become more technologically advanced. Fighting this threat, we are told, requires more innovative technologies, extending their remote-sensor tentacles near and far to preempt risky bodies en masse before and after they reach the US border as well as the more than seventy other militarized borders around the world.[8]

In fact, AVATAR may one day replace the Border Patrol, with some journalists asserting its superiority over humans, since "AVATAR doesn't see race, doesn't play favorites and doesn't get tired."[9] Although activists claim that Arizona engages in racial profiling, authorized by Arizona's SB 1070, machines ostensibly cannot *see* race, and computers detect only universal patterns that calculate an individual's levels of threat versus trustworthiness. Thus, one of the questions I follow here is, During a posthuman era, what are the consequences of automated technologies that detect and racialize *suspect life* under the guise of scientific neutrality and automated algorithms without human interference?

Suspect life here refers to the racial bias preprogrammed into algorithms that compute danger or risk from certain human movements in regions such as border zones. Legislators approve funding for biometric surveillance in the search for more efficient and unbiased techniques for identification.[10] Border biometrics promise a progressive liberal future where ubiquitous surveillance frees us from unknowability and a fall back into a primitive state. Claims to objective science and technological development conceal the broader history of war and genocide against Indians (and Mexicans thought to be Indians) and colonial divisions of race and territory. In addition, these claims to objectivity contradict evidence on the ground that the majority of deportees are Latinx families, many of whom are deported without the benefit of a court hearing.[11] Attending to the ways surveillance technologies are developed and put into use reveals the novel forms and continuing legacies of racial science and the histories of colonialism, war, enslavement, and empire.

The Border Biosecurity-Industrial Complex

Justin Akers Chacón and Mike Davis refer to a "border-industrial complex" to emphasize how border enforcement has become "a privatized and profitable enterprise" orchestrated by the DHS to increase border control funding, detentions, and interior investigations.[12] Arizona leads this biosecurity complex as the state with the largest numbers of incoming unauthorized migrants following the walled closure of borders in California and Texas. Funneling migrants through the blazing hot desert has also spiked the death toll in this region, leading longtime sociologist of immigration Douglas Massey (and then President Donald Trump) to call Arizona "ground zero in the war on immigrants."[13]

Desperate for more local power to criminalize immigrants, Arizona passed the first identity theft law in 1996, the same year President Bill Clinton passed the Illegal Immigrant Reform and Immigrant Responsibility Act (IIRIRA). The law extended unprecedented power to states and local police to enforce immigration law, especially as identity fraud became a criminal offense, which criminalized false claims to citizenship and rendered undocumented immigrants with minor misdemeanors eligible for deportation.[14] Motivated by Arizona's inability to control its own borders owing to the political inaction of the federal government, IIRIRA increased funding for Border Patrol agents and surveillance technologies, called for expansion of the border fence, fortified the interior enforcement of immigration laws, and mandated the use of biometrics on ID cards by 2000.

Since the act's passage, border security spending swelled to $18.5 billion a year in 2013 and continues to increase.[15] Contributing to this surge in spending were a spate of DHS-backed plans to increase new technologies and information systems along the border, including the Southwest Border Technology Program (2001/2014) and the 2011 Arizona Technology Plan. These propositions called for the deployment of radar, sensors, and cameras first in Arizona and then along the entire southwestern border. President Trump continued this trend by investing billions of dollars more toward a physical and virtual border wall.[16] In 2018 the US Customs and Border Patrol's budget reached $16.3 billion. It is the largest federal law enforcement agency under the DHS, and its mission extends far beyond the US border.[17]

Arizona has taken advantage of this funding gold mine by resculpting its economy into a hub for high-tech border and aero-defense industries, while expanding jobs in its penal economy to control immigrants crossing the border. The hypercriminalization of migrants strips them of due process as they are fast-tracked into detention facilities, often devoid of legal oversight or protections, where they face indeterminate sentences. With the aid of IIRIRA, the number of deportees nearly doubled, from 70,000 in 1996 to 114,432 in 1997.[18] Today identity theft constitutes an aggravated felony that can lead to an automatic jail sentence of up to twenty years, or deportation without the possibility of return for up to twenty years.

The DHS found Arizona an ideal home for border security, given the state's status as a critical border-crossing zone and a laboratory for new surveillance technologies, as well as its draconian anti-immigration policies, such as SB 1070 (2010). Popularly known as "stop and frisk," this law allows officers to request identification papers from anyone who raises a reasonable suspicion of illegal presence, a practice that many argue buttresses racial profiling. With one of the largest federal budgets for intelligence gathering and a research arm focused on science and technological development, the DHS funds twelve university research centers tasked with apprehending hidden threats such as terrorists, migrants, diseases, and microbes that could attack the nation. One of these DHS Centers of Excellence—the University of Arizona's BORDERS: National Center for Border Security and Immigration—states its mission as "developing innovative techniques, proficient processes and effective policies to help meet the challenges of border security and immigration, with a specific focus on screening and deception detection."[19] To accomplish this goal, AVATAR is intended to assist border officials with faster and more accurate detection of deception.[20]

Within this broader context of identity fraud and suspicion, the ID document has morphed from a textual representation of the self (the document *represents* me) to one that relies on biometrics to verify and stabilize the "truth" of the body (the document *is* me).[21] As biometrics increasingly augment, and at times replace, paper documentation, these technologies recycle the settler colonial racial science of biological categories of race and the human found in nineteenth-century anthropometry such that today "bones, voices, DNA, and fingerprints serve as identity documents" and even a unique signature.[22] The body-as-data converts biology and life itself into transferable bits of code. While scientists revered body parts (such as skulls) in the nineteenth century as empirical proof of racial, and especially mental, difference and as clues to unlock the past, today biometrics are imagined as alienated from ancestry and race, as they typify distinct (but also universal) traits unique to each person and thus form the signatures of a computer-readable identity footprint to be deposited into a database.[23]

By turning the body from a subject who speaks into biological matter, biometric technologies dehumanize their targets, even as the unconscious signs of the body present a supposedly more scientifically verifiable truth than human verbal testimony. Since the perception is that humans deceive and avert apprehension by hiding behind false statements and documents, biometric technologies shift border control from the subjective interview to the automated body, and from legal adjudication to algorithmic governance.[24] Border security relies less on the narrative interview (or confessional) than on the body-as-matter, shifting the focus from subjects defined by narrative and biographical information to the body's preverbal status. I say *preverbal* because biometrics convert the body into raw data in a way that hearkens back to the racial science of the nineteenth century, when the body was presumed to communicate emotional cues that erupted in the unconscious or unintentional microgestures of the body.

The anthropologist Marcello Levi Bianchini developed a kinesiological biotypology of race in the early twentieth century that separated "primitive" from "advanced" races, arguing, "In primitive societies all the affective reactions of the individual and collective psyche are essentially manifested by phenomena of movement."[25] For Bianchini, primitive societies displayed a wider range of motor expressions (or affects) than the more evolved societies. Around the same time, the French physician Félix Regnault turned to film to document that those he called savages had no language and instead spoke through gesticulations of the body, that their "savage locomotion" was closer to nature, untouched by cultural influences.[26] As discussed in chapter 1, nineteenth-century ethnologists, anthropologists, linguists, and others categorized Indigenous gestural language as a prelinguistic form of communication defined as overly emotional, feminine, and not-quite-human, more akin to the communication of animals, the mute, and those with disabilities. Given that language differentiates human from nonhuman, gestures encapsulated Native Americans as a primitive form of humanity. Similarly, AVATAR's turn to the body's physiology preprograms immigrant flesh as a primitive threat to the civilizational order of the US nation-state.

Attempts to visually produce evidence of the racialized body as a criminal threat or vector of disease at the border can be seen in the rise of technologies used to identify a range of deviants. Police in France and Britain used photography in the 1840s to capture the face, or ideal type, of the criminal.[27] The search for a more secure mark on the body, such as a fingerprint, to govern bodies also emerged in the British colonies of India during the Sepoy Mutiny in 1857–59. The British needed to better identify and track natives not only during this time of rebellion but also when British rulers feared rampant forgery and theft by locals. Back on the US-Mexico border, racial fluidity has long generated the desire for better documentation; immigration historian Erika Lee details how some Chinese donned Mexican clothing to pass the border undetected.[28] Further, Chinese and Mexicans—thought to be "unhealthy" or "unhygienic"—were cast as a danger to the nation and thus barred entry.[29] In fact, Mexican migrants in the early twentieth century were quarantined, sprayed with DDT to kill any germs or diseases, and branded to mark them as authorized.[30]

From the development of biometric fingerprints in India to the tracking of Native American and Mexican footprints at the US-Mexico border to current forms of algorithmic governance, the state contorts the body into an "authoritative text" of future behavior. Whereas the first stage of security depended on the control of documents, in this second stage the body becomes digital data that speak their own algorithmic truth and is presented as more accurate evidence than a person's spoken claims. For example, as narratives of political persecution are treated as suspect, asylum law in France has turned to the (suffering) body for more authentic claims, leading to what Didier Fassin calls *biolegitimacy*.[31]

Simone Browne argues that the state's biopolitical project cannot be severed from violent sentiments of racism or anti-Blackness. She foregrounds surveillance as an ongoing project from slavery to the present that aims toward "the imposition of race on the body," a process she theorizes as "digital epidermal-ization."[32] Yet, as I unpack here, historical specificity is critical when assessing the kinds of racial evidence deployed by surveillance technologies and tactics such as AVATAR. For example, while the "stop and frisk" techniques authorized by Arizona SB 1070 were vehemently attacked by activists and organizations as racist in their targeting of brown and Black bodies, AVATAR sensors go beyond skin color to mobilize a seemingly postracial scientific methodology to appre-hend those with suspect physiology.

Today governments not only track (demographic) populations based on predictable patterns of life and death but rather seek assessments of actuarial risk, a concept produced at the intersection of the life sciences and rapid com-putational algorithms that evaluate risk through biometric patterns that can purportedly be detected in advance of deviant acts. As the life sciences meet artificial intelligence, intelligent design and consciousness have moved into the body's behavioral movements rather than stemming from the conscious narration of the brain, alienating the active properties of the body-as-matter from human intelligence, consciousness, intention, and agency.

Today the so-called terrorist is presumed to follow behavioral laws of nature that resemble automated life, or mechanization, offering science visible and predictable patterns of movement. As articulated in a statement by the Inter-national Association of Chiefs of Police, suicide bombers are reputed to dis-play "an unusual gait, especially a robotic walk. This could indicate someone forcing or willing himself or herself to go through a mission."[33] Important to note here is the idea that suspicious gait may turn the body itself into a primi-tive weapon. The racialized body has the potential for being weaponized as a technology of terrorist warfare (a bomb); it is racialized as primitive owing to its association as an object that can be (remote-)controlled by a force larger than the self-determined, democratic individual. While an extreme example, this robotic body becomes racially suspect through its association with mass destruc-tion, or a racial biomass, a body tethered to its web of connections, a mass not driven by individual rationality, the brain, or the self but dangerously susceptible to the higher call of religion, culture, instinct, and the (totalitarian) collective.

Technologies for seeing danger in war zones view spectacles of bodies-as-mass through typologies of race, sexuality, chaos, and abnormality. Despite claims by the Department of Defense that the military and intelligence now target individuals, the surveillant gaze is already programmed to see threat on bodies crossing the border, a racialized geopolitical zone categorized as a hot spot, or a region long associated with imminent threat.[34] The US-Mexico border has long been viewed as a zone of (sexual) excess and danger, depicted especially as a place where hordes of migrants infiltrate the nation.[35] The "hot spot" metaphor recycles epic battles between nature and technology through its connotation of an excess of heat, affect, technological connectivity, disease, illicit commerce, and erratic behavior, as individual bodies become racialized through the process of massification.[36] Massification entails a process of classification through which space and bodies are coded as threats owing to the fear that they will spread danger through biological and sexual contamination, emotions, communication, and/or interactions with others. Seeing threats at a region designated as a hot spot blurs the borders between an individual and mass danger. *Hot spot* has a range of definitions: a region with a relatively hot temperature in comparison to its surroundings (geologically, as volcanic activity); a place of significant activity or danger; an infectious zone of the body; and a zone of commerce lacking security.[37] As we saw in chapter 2, newspaper articles characterized the Tohono O'odham reservation's "border" as a black hole or security gap in order to justify the urgent need for border security infrastructure. Erased from view are those O'odham whose land extends beyond this artificial line that severs one nation from another. What unites these definitions is the ways border optics see bodies through exceptional danger that triggers a red heat print of thermal imaging activity on the computer screen. It should not surprise us that, since 2001, border patrol personnel scan the border using infrared thermal imaging cameras that see humans and animals through heat imprints, especially at night and in bad weather.[38]

From War to the Border: Deception Research

Given that border surveillance has become a lucrative global venture-capital market, I interviewed staff at the University of Arizona's BORDERS Center and Tech Park, a sprawling desert facility where faculty research is tested, patented, and sold to military defense companies, who adapt their wares to border security. In March 2015 I scheduled an interview with Elena, one of the key managers and research developers at BORDERS, during a visit to the University of Arizona in Tucson. Elena helped build a consortium of faculty from eighteen universities that prepared a successful proposal for the DHS Center of Excellence grant for "Border, Security, and Immigration."[39] With an advisory board of two retired military officers, someone from a security investment firm, and two others from the Immigration and Naturalization Service, the center supports faculty, graduate, and undergraduate research on border security. The

DHS funds projects that meet its approval, such as AVATAR, as well as a project by aerospace mechanical engineering students researching locust wings, which they hope will lead to minidrones deployed at the border.

The faculty most interested in participating in BORDERS research projects also wield credentials in military research and corporate funding. Two primary faculty researchers of the AVATAR project—Jay Nunamaker, a professor of computer engineering and communication, and Judee Burgoon, a professor of psychology and the communication of deception—received Department of Defense funding before joining BORDERS. Nunamaker is also affiliated with the University of Arizona's Deception Detection Lab (where some of AVATAR's technologies were developed), sponsored by the Air Force Office of Scientific Research and the Army Research Institute. And for faculty who may be reluctant to support a project like AVATAR, their students may pitch their projects to security industries such as the Department of Homeland Security since they offer lucrative careers for minority students at universities that qualify as minority serving institutions or Hispanic serving institutions, like the University of Arizona. The DHS also offers internships and research jobs affiliated with the fields of homeland security science, technology, engineering, and mathematics. In fact, faculty positions in homeland security disciplines are on the rise across many college campuses.[40]

Burgoon publishes prolifically on how deception can be communicated and detected through the body. Her scholarship is heavily influenced by Paul Ekman's research on deceit, in which he argues that since lying can be detected in emotional cues that erupt in the body, liars' bodies betray them.[41] Ekman trained Federal Bureau of Investigation and Transportation Security Administration personnel and doctors to detect lies through a person's "micro expressions," as found in their body language, voice, and facial movement.[42] His research originated in the observation of psychiatric patients whose speech could not be trusted, especially since some would lie or "simulate feelings of optimism" in order to be released early from the hospital.[43] Later he was asked by the Federal Bureau of Investigation to test his skills on airport passengers, who might fraudulently pass security surveillance by posing as citizens with the aid of false documents. Ekman's research departs from the work of Sigmund Freud, who detected the unconscious through slips of the tongue, or speech. Ekman and (later) Burgoon work from the premise that the hidden signs of deceit are "emotional leakages" lodged not simply in speech but within the unconscious fibers of the body.[44] For scholars influenced by Ekman's research, the flesh operates as a confession, and the body's involuntary movements, gestures, and physiological responses (rather than thoughts, desires, and imaginings) are visible and verifiable signs of arousal, guilt, anxiety, or fear prompted by deceitful communication.[45]

In his 1913 book *Totem and Taboo*, Freud argued that modern forms of socialization, especially behaviors of conformity, are shaped by shared primitive origins. The unconscious was the dark place of animalistic instincts within ourselves, outside rationality and conscious intelligence or thought, where traces of primitive impulses and repressed sexual desires lingered, including irrational fears or perceived threats to our existence. His theory relied on evidence from the Aborigines of Australia about their primitive use of totems, or social taboos, to control behavior. For my purposes, it is helpful to consider Freud's three stages of human evolution from primitive to modern. Belief in animism, from the perspective of Indigenous peoples, was the earliest stage, then religion and, at the modern apex, science. He contends that animism is the primitive belief that every object in the universe possesses a soul. In other words, for Freud, this overblown sense of humans' control in the world, in which tribes saw each human action as having a ripple effect on other objects across the universe, was at the root of modern narcissism. This led him to the conclusion that Aboriginal belief in the inseparability of human-animal-object relations led to an obsession with magic, or control over human behaviors (since they perceived these actions as having the power to negatively alter the world around them). He related Aboriginal social control over individual behavior with religious repression and intolerance. In this way, Freud pitted primitive magic and religious conformity against the scientific method—in his case, psychoanalysis. Psychoanalysis was suited to eradicate the unknown, or lingering primitive beliefs, and catapult society into an era of social scientific harmony and freedom.

Ekman and Burgoon both departed from Charles Darwin's scientific theory that emotional expressions in humans are similar to animals and thus innate, a vestige of our animal past. Instead, they sought to situate emotions within a learned behavioral context such as communication to argue that emotional reactions are both biologically universal and socially learned.[46] While Ekman trained people to see deceit on the body, Burgoon argues that the cues of deceit are almost impossible for humans to decipher.[47] Thus, for Burgoon, because psychological emotions erupt through behavioral cues found in the body's automatic physiological movements, these automated signals are best detected by computers.

Burgoon teamed up with Nunamaker, a BORDERS center scholar in the fields of information management, computer engineering, and communication who has generated over $100 million in grants to the university. Burgoon and Nunamaker were interested in how language expresses a behavior that communicates guilty or innocent emotions: *what* was communicated was less important than *how* communication betrayed those who lied. Algorithms, they claimed, offered better data detection than humans, whose accuracy declines with fatigue. For example, in a paper on how to improve deception detection, Burgoon, Nunamaker, and Brent Langhals hearken back to Cold War research that found that humans fail over time as vigilant observers of insurgents who smuggle weapons into the United States. Today, they argue, "the specified stimuli are the presentation of hidden 'weapons' within a given data stream."[48] In line with the need to apprehend threat and risk before they cause damage, evidence of criminal intent morphed from (1) proof of premeditated action to (2) a weapon hidden on the body to (3) the body's unconscious testimony, as data automatically computed by algorithms.[49]

Nunamaker's interest in deception began during the 1990s with research he conducted at IBM and for the US Army and Air Force. Fascinated with the ways deceitful employees betray themselves, he differentiated lies by "looking for a statistical prevalence of evasive language and 'hedging words.'"[50] When Nunamaker met Burgoon in the 2000s at the University of Arizona, he encouraged her to automate her tedious methods of gathering data on linguistic changes in tone and word usage as well as body movement (requiring hundreds of hours spent recording microscopic moves from a video screen). They set out to develop a machine that could rapidly collect this data. Thus, rather than train individuals to detect lies, they built a lie-detector machine and then, more recently, helped create AVATAR based on Burgoon's psychology research and Nunamaker's experience in computer engineering and mass computer networking platforms.

Burgoon and Nunamaker received Department of Defense funding for ten years to work on a previous iteration of AVATAR. In 2006 the army's Polygraph School (renamed the Defense Academy for Credibility Assessment) funded them to develop a noninvasive, remote lie detector for military applications and interrogations. While the old polygraph lie detector was encumbered by straps that had to be manually attached to the body to monitor heart rate, Ekman and Burgoon assert that AVATAR's remote sensors will more accurately detect aberrant vital signs.[51] The first polygraph lie detector was created in 1921 by August Vollmer, popularly known as the "father of modern policing," who developed a range of scientific breakthroughs, or gadgets, to make crime control more efficient, leading him to establish the School of Criminology at the University of California, Berkeley. Steeped in the evolutionary racial typologies of his time as well as anthropometrics and biometrics, Vollmer hired a physiologist at Berkeley to design the country's first lie-detector apparatus in an attempt to assess criminal guilt.[52]

Despite his background in polygraphs, Nunamaker has declared that AVATAR "isn't a lie detector. We're looking for signs of risk."[53] Each person who approaches AVATAR must scan their ID and then answer questions posed by the virtual border official on the monitor screen: Is this document yours? Have you ever committed a crime? Are you carrying a bomb? These incriminating questions aim to verify the "truth" of the narrated responses as AVATAR rapidly collects and stores data on a person's physiological and linguistic signals. Especially appealing to researchers are AVATAR's sensors, which can monitor the body's *involuntary* movements, such as the cardiovascular sensor. It tracks movements such as trembling hands as well as spikes in blood pressure, heart rate, and body temperature while recording facial twitches, pupil dilation, and eye gaze. Churning these data through algorithms, AVATAR separates border crossers into two types: those who can be safely admitted (relaxed body movements, normal behavior, and stable vital signs characterize those with seemingly nothing to fear) and those who are potentially a threat (elevated blood pressure or other markers of stress register the potential for lying, guilt, fear, irrationality, and insecurity). Travelers who show signs of distress on the data collected by sensors are required to undergo further scrutiny and questioning by a human officer.

From the University to Corporate and Military Research and Development

The DHS's academic disciplines and funding structures promote science and technology as the solution to social and environmental problems at the border. The technological products funded by the DHS are evaluated through an assessment process called *design science*, which checks their practical applicability for use in the marketplace. The fields of science, technology, engineering, and mathematics more broadly contribute to a trend in how universities produce knowledge, moving from theorizing and explaining the existing world to shaping it with innovative objects. In addition, new patent laws have altered the nature of faculty research. In 1980 the Bayh-Dole Act granted more intellectual property rights to faculty (rather than the university), who now benefit from collaborations with companies. Once a product is funded, the companies take 51 percent of the profit, leaving the faculty and the university to divide the remaining 49 percent.[54]

The University of Arizona's vice president for economic development, Bruce Wright, acknowledges the synergy between the university and commercial development: "We have seen an increased focus among companies specializing in technologies related to border security. The BORDERS center will assist in the development and commercialization of new technologies and policies that address issues of border security and immigration with the hope of creating new companies utilizing technologies developed in the center."[55] The sensors used in AVATAR were individually developed and patented across eighteen universities and are continually adapted to expand the product's use value and profitability. For example, while its remote sensors were first tested at actual checkpoints in Nogales, Arizona, to differentiate safe border crossers from dangerous ones, the AVATAR tested in Singapore could speak seven languages. With scant resources, Singapore's AVATAR could accomplish what most humans cannot, monitoring a multilingual border zone crossed by more than thirty thousand people per day.[56] Given this success, AVATAR will be implemented at the most congested border checkpoints, such as the US-Mexico border at Nogales and the Singapore-Malaysia border. In Nogales a tall metal wall ends abruptly at a multiple-lane checkpoint entry where one can see cargo trucks waiting in endless curvy lines to the right and vehicles snaked in multiple lanes to the left. The wall here is reminiscent of a prison or a concentration camp, with razor wire spiraling along the top. Just as striking is the Singapore-Malaysia checkpoint, isolated on a causeway surrounded by water. Here, too, lines of cars wait far down the peninsula for their turn to be verified by border officials.

While it is still unclear how many purposes and locations AVATAR has served since its release in August 2018, Elena, the spokesperson for BORDERS, expressed enthusiasm for AVATAR's potential applications within a global marketplace during my interview with her in 2015. Its lasers, she explained, were developed by the University of Washington for medical purposes, such as taking the vitals of patients who lie about their alcohol and drug intake, thwarting doctors' ability to successfully treat them.[57] These same lasers will be used for border control, including the detection of drug smugglers and trafficked persons. The air force also wants this technology to monitor and transmit soldiers' vital signs on the battlefield. Clearly, the value of technological development and research is tied to its perceived social utility for war, medicine, and border control, with less attention to how the context of war shapes what we see as a threat.[58] Yet these developments have not escaped public resistance. Protests have ensued in Arizona demanding a demilitarization of the borderlands, and communities are popping up to protest the use of mobile surveillance vans. For instance, protests erupted in New York against unmarked vans armed with mobile X-rays that scanned buildings and people unannounced, as cops patrolled suspect regions from a distance.[59]

Unfortunately, obscene profits are made in these industries, leading many universities to seek out ways to tap into these revenue streams. The University of Arizona purchased Tech Park from IBM in 1994 to build relations between the university and tech industries. As a testing zone for global border-security technology that is located sixty miles from the border, Tech Park showcases over forty border-security companies. On my tour of the site, Betty, its public relations representative, drove me to see the most lucrative surveillance tower project, contracted to Elbit Systems of America, Israel's largest private military manufacturer, to replace the defunct IBM towers designed by Boeing. The operational failure of these eighty-foot Boeing towers in 2006—costing taxpayers $7.6 billion—catapulted the DHS into media scrutiny, contributing to the DHS's current demand that products be tested before any contract is granted.

Elbit Systems of America secured a contract with the DHS in 2014 based on the construction of surveillance towers called *integrated fixed towers*, which included a plan to build the towers along the Arizona-Mexico border (see chapter 2). These 160-foot-tall surveillance towers are equipped with high-definition cameras (that can peer into homes and through walls), night vision detectors, sensors, and radar and have the ability to send data to Border Patrol agents far from their actual location. As a "force multiplier," integrated fixed towers positioned along the border allow one agent to surveil a broad region of the border that it might take a hundred agents to patrol on foot. Sitting in a safe office building in Ajo, Arizona, for example, an agent surveils the borders of California and Arizona remotely and transmits the exact location of roving border crossers to agents waiting to be dispatched.[60]

Arizona's border-surveillance techniques and the logic behind them are imported from Israel's "laboratory of war" in the Occupied Territories of Palestine.[61] According to Naomi Klein, 9/11 offered Israel the chance to rebrand itself as a showroom of military technology, "turn[ing] endless war into a brand asset."[62] More than four hundred Israeli companies export security-related products, with a turnover of $4.5 billion for 2007; thus, Israel climbed out of its 2000–2003 economic slump by investing in a profitable high-tech security market. Industry enthusiasts pitch their wares as tested and proven in the real-life context of war against Palestinians.[63]

Elbit advertises its homeland security systems to an international community, and it was selected by the DHS for having over ten years of experience "securing the world's most challenging borders," including the installation of smart fences in the West Bank and the Golan Heights.[64] Elbit was also the first company to test drones on the US-Mexico border and more recently has sent its drones to patrol the Mediterranean Sea based on a European Union contract aimed at controlling migrants from northern Africa. As journalists Todd Miller and Gabriel Schivone note, "Like the Gaza Strip for the Israelis, the US borderlands, dubbed a 'constitution-free zone' by the ACLU, are becoming a vast open-air laboratory for tech companies."[65] Taking their cues from Israel, influential investors and venture capitalists also hope to turn Arizona's sluggish economy into a high-tech hub, exacerbating gentrification and forcing many unemployed workers into penal labor markets. In fact, Arizona's governor sent delegations of lawmakers and technology experts to Israel to research whether the laws and technologies that strip Palestinians of land, mobility, and legal status could be deployed in Arizona. At the same time, a youth delegation of O'odham traveled to Palestine—invited by Stop the Wall, the Palestinian coalition that opposes Israeli walls and surveillance—to share information and strategies that unite these territories and people against military occupation.[66]

Elbit's underground sensors are being tested for detecting aberrant changes across the landscape, differentiating cattle from humans, or trains from other vehicle vibrations setting off the alert system. Both AVATAR and Elbit's virtual wall incorporate smart sensors inspired by the analytics software called AI (Artificial Intelligence) Sight, developed in Israel in collaboration with partners in the United States: "From this data, the technology determines what is typical—and sends a proactive alert on what is out of the ordinary."[67] According to AI Sight, anomalies are detectable only through the computation of big data sets that track statistical variation, or behavioral statistics that deviate from the norm. Smart computers accumulate information from real-time movements and compare it with a historical analysis of a place to chart how patterns of behavior evolve, allowing the system to not only recognize norms but learn new ones, adjusting to anomalies across time. Thus, behavioral pattern recognition is not limited, they boast, by predefined behaviors. These "smart" systems learn to interpret the usual movements occurring at a spatial location, while documenting unusual activity as suspicious until it becomes standard and then is reintegrated into the normal functioning of a place. In this framework, suspicious activity is anything that deviates from the norm, while there is little interest in defining which bodies and what behavior the norm relies on.

The scientific methods behind AVATAR include what the company Israeli Homeland Security calls its "Suspect Detection Systems" (SDS), a combination of lie detectors, interrogation techniques, biometric data sensors, and algorithms used to assess the hidden "hostile intent" of travelers at the Israel-Palestine border and at US borders. The Israeli SDS does not test for general nervousness but rather looks at patterns of data taken from Palestinians, Israelis, Americans, and other groups in a single area to determine which body behaviors stand out from the rest.[68] Despite Israeli Homeland Security's promotion of SDS as objective, and thus transportable to any location, suspicious behavioral patterns are coded through Palestinian movements and thus severed from the context of Israeli military terror—the takeover of territory, war, and the security apparatus of border checkpoints. Given the intentional maiming of Palestinians through violence and war, and the curtailing of their mobility and access to water, agricultural lands, and livelihood, it is plausible that the very same disabilities violently produced through the war with Israel would constitute a suspicious sign or footprint of deviation from the norm.[69]

Along the US-Mexico border, the wall itself is weaponized in various places to cripple or even kill those who attempt to cross it. Sharp metal sheets along the wall and razor and glass mesh wire are designed to sever body parts and gash the flesh, causing many to fall down onto hard concrete.[70] And as many scholars have argued, the state's funneling of migrants into swelteringly hot desert regions also weaponizes the natural terrain as thousands have suffered severe injuries, illness, and even death.[71] Promoting SDS as a tool to capture terrorists, industry advocates promise that it can assess potential threats within seconds to prevent incidents before they are carried out. This technology supposedly discerns and disarms the hidden "hostile intent" of assailants before they commit their intended acts—especially, advocates claim, "since [the] 9/11 terrorists came to the US with full documents, and no weapons, only intent."[72] Asserting that the technology is based on "Behavior Pattern Recognition without Human Interference," Israeli Homeland Security advertise the superiority of artificial intelligence over humans in the battle to rapidly calculate anomalous behaviors across time and space.[73]

In 2015 another media scandal hit the DHS, causing it to change its airport security program SPOT (Screening of Passengers by Observation Techniques) to FAST (Future Attribute Screening Technology), which is based on some of AVATAR's remote sensors. Starting in 2007, airports across the United States and in Romania had trained personnel in the supposed science of seeing suspicious behavior through SPOT, a protocol developed by police at Logan Airport in Boston that relies on the lie-detection research popularized by Paul Ekman. Ekman was consulted for his research from the 1970s on the Facial Action Coding System, which cataloged hundreds of expressions of emotions.[74] Over 160 agents were dispatched to airports around the country, armed to see suspicious feelings and behaviors on travelers' faces and bodies, especially emotions that people might try to conceal.[75] Suspicious behavior included unconscious acts such as fidgeting, clearing one's throat, staring at one's feet, and engaging in excessive grooming.

Public outcry followed a 2015 report by the *Intercept* revealing that the program was racially biased to target immigrants, even though antiterrorism rhetoric was deployed to justify the expenditure of public resources. The government spent about $900 million (from 2007 to 2015) to identify terrorists at airports but detained only undocumented migrants. Of 429 passengers identified for a second screening, 14 were detained for being in the country illegally.[76] After this criticism, the program moved to a remote testing phase in which actors were videotaped and then examined for suspicious movements that would confirm they should be detained for being terrorists. In its testing phase, FAST, similar to AVATAR, aimed to identify individuals exhibiting physiological indicators associated with "malintent" or "suspicious behavior indications" as proof of their intent to cause harm (especially intent within what is called a "future time horizon").[77] It does so by analyzing "specific psychophysiological signals and behavioral attributes, e.g., respiration, cardiovascular response, eye movement, thermal measures, and gross body movement of a screened individual."[78] According to a memo authored by the science and technology director of the DHS, the consequences of having one's biometrics flagged by FAST at airports and borders are varied and "can range from none to being temporarily detained to deportation, prison, or death."[79] These potentially severe consequences would be for someone identified as suspicious who was then sent to the proper authorities for further questioning and assessment. Surprisingly, the DHS report provides no further explanation about what evidence is needed for someone to face death when found guilty of committing a crime in the future.

In light of the failures of the clunky and expensive Boeing surveillance towers that triggered public concern over privacy and fears of totalitarian government control through the all-pervasive gaze of surveillance, industry specialists are moving into less visible and more remote technologies.[80] The smart technology driving AVATAR represents just this move to border-surveillance tools that bypass human vision and touch and thus fail to register as (sexually) invasive, racially biased, or malicious. For this reason, Anduril Industries is currently developing what journalist Steven Levy calls "a digital wall that is not a barrier so much as a web of all-seeing eyes, with intelligence to know what it sees" to detect unauthorized border crossers from Mexico. Palmer Luckey, a young entrepreneur who sold his virtual-reality software to Facebook, pitched his virtual-reality headsets to the DHS as tools that digitize movements across the landscape. As is common knowledge in Israel, a virtual surveillance wall must accompany a physical border wall to improve the speed of apprehension. Luckey's digital wall gathers data from any sensor along the border, promising a future of "geographic near-omniscience."[81]

Many of the visual technologies developed at the border to see suspect life remotely derive from those designed for war. Gait recognition uses infrared cameras developed for covert military operations to interpret movements from afar, patterns of life that may be so small or elongated that they would pass undetected by the human eye. In Baghdad the US military uses the Unblinking Eye, a 24/7 multisensor surveillance device, to detect anomalies from the safety of an unmanned aircraft high above, taking "a visual signature of how the target walk[s], travel[s] in groups, or engage[s] other people. The ability to recognize a target's gait, dress, companions, parking patterns, and so forth bec[o]me[s] high-confidence targeting indicators because of the hours of pattern of life observation."[82] Implicated by the portrait of patterned movement, the body becomes a "target," echoing Rey Chow's argument about terrain: "In the age of bombing, the world has also been transformed into—is essentially conceived and grasped as—a target. To conceive of the world as a target is to conceive of it as an object to be destroyed."[83] Once remote technologies from war are transferred to the border, humans are identified as visual targets, already knowable on the screen as potential enemies and threatening others. The border has become a virtual battleground where visualizing the threat is part of the everyday targeting of bodies for the project of securitization, or preemptive combat.

Aerial viewpoints are normalized as the tools for envisioning bodies-on-the-move as abnormalities, threats that must be externalized or deterritorialized—cut out before they can take root in social worlds that hide their movement within a labyrinth of connections across time and space. In the case of Unblinking Eye, the human target is placed within a network of movement (drawn from observations of neighbors visited, GPS monitoring, and tracking of online purchases and emails) for a more accurate portrait of aberrant, or dangerous, life. Patterns of life, even online activity, deemed as having the potential to "incite violence" can lead to disciplinary and legal action against those already racialized as suspect.[84] As this technology moves from war to the border to everyday life, it is crucial to examine what kinds of lives, consumption patterns, and movements become suspect, and what radius of influence then implicates others.

Ostensibly armed with the power to detect an individual's risk of future suspicious behavior, algorithmic data trump human perception and support claims that biometric identification is "the purest and most non-discriminatory form of personal identification."[85] As one industry representative states, "Race, sex, and age are not generally considered or factored into the mathematics of a facial recognition algorithm. These aspects are largely biographic and contextual data descriptors. . . . [I]t is in the best interest of our industry to develop highly accurate algorithms that do not consider such aspects at the algorithm level."[86] These claims to mathematical objectivity fail to consider the militarized force of data, which can now serve as a weapon of social control, a means of penal adjudication, and a tool for the sovereign control of bodies and territory. The promise of algorithmic objectivity is vigorously contested by those who say that racial profiling distorts them, especially in criminal cases.[87] In addition, statistician and mathematician Cathy O'Neil asserts that automated algorithms discriminate based on neighborhood (zones designated as high risk are often regions where poor people of color live), past behavior, and physical appearance.[88] They measure success through increases in apprehensions for crime (even though the higher police presence no doubt leads to more people locked up in jail for petty crime), efficiency, and profits rather than equality and justice. And because everyday people are often outside the logic and understanding of how algorithms differentiate between risk versus safety, these automated computers wield the power to make those decisions for us.

Conclusion

The turn to remote identification and surveillance makes it apparent that many do not have the right to their own bodies, neither to the information collected on them nor to the interpretation produced by computer algorithms—what Jasbir Puar labels *computational sovereignty*.[89] The fifty invisible scanners of AVATAR invade the body remotely and mine it for data. The body becomes open territory that can be colonized by those who claim expert knowledge of its meanings while asserting the right to sovereign control over its mobility and future possibilities. I have located 9/11 and the fear of document fraud as significant in triggering another colonial regime of security based on the racial fear of primitive locomotion, detectable only in the secret recesses of the body's silent footprints. This primitive locomotion codes the ways the state sees deception through the Native threat at the border.

Biometric technologies recycle a settler colonial science of the body that sediments an individual's purported criminal propensity into a national database, with lasting consequences. If a person's life was traditionally recorded through a historical narrative tied to human time and space, today it is churned into data bits that record micromovements of suspect flesh imperceptible to humans. Once entered into databases, these snapshots of life are severed from behavioral actions and universalized to become an aspect of identity tied to us for an undetermined duration, regardless of when and where one is visually snagged by the surveillant gaze. In fact, while old offenses are deleted from citizens' records after seven years, the 1996 Antiterrorism and Effective Death Penalty Act stipulates that any crimes committed by noncitizens postentry, regardless of the severity of the crime, will remain permanently on their records and thus can lead to prison, detention, and deportation.[90] Once in the database, an individual's biometric details can be matched to future criminal activity, amplifying the chances that Black, Latinx, and Native people—those overwhelmingly caught in the dragnet of data collection at prisons, in sex offender lists, and as welfare recipients—will become suspects for crimes that have not yet been committed. The danger of escalating the computational risk of racialized male bodies is most apparent in the cases of police violence against mostly male African Americans and Indigenous and Latinx migrants who are shot without proof of aggression, many of whom are also disabled or suffering from mental illness. And border patrol officers were deployed to join federal police to suppress protests against widespread police shootings of African Americans such as George Floyd, Breonna Taylor, and others. Customs and Border Patrol "even flew an unarmed Predator surveillance drone over Minneapolis."[91]

For immigrants, once identity theft became a criminal and then a federal offense, lacking proper documents or holding fake ones became evidence of their desire to deceive and commit harm, a legal quagmire that has caused thousands of Latinx workers to be deported without hearings. While immigrants are supposed to have the right to due process, a 1996 statute permitted authorities to deport them without a hearing, a lawyer, or the right of appeal, a process known as *expedited removal*. Currently, expedited removals apply to undocumented migrants who are found within one hundred miles of the border within fourteen days of entering the country. There are many exceptions to this rule that allow this geographic limit to be exceeded and that permit expedited removals up to two years after a migrant has entered the country. These exceptions are what the Trump administration wanted to make standard, including his expulsion of two hundred thousand asylum seekers without a hearing based on their potential to spread the coronavirus, even though reports show migrants were infected with COVID-19 in detention centers and in transit during deportation.[92] In fact, Palantir, the same company that is building the virtual border wall, creates software for ICE and for Health and Human Services.

It is not a stretch to see the potential for Palantir's COVID-tracking system (called Protect Now) on immigrants, Native peoples, prisoners, and so on.[93]

Rather than simply demanding more legal protections, immigrant and Native American activists align themselves with forces more intelligent and powerful than the state by relying on migrant testimony and stories of human-animal-plant responsibilities to each other's thriving to offer fuller accounts of human life.[94] Activist slogans such as "No One Is Illegal" and art images of the monarch butterfly assert the sovereignty of nature by extending the right of migratory flight to all earth's inhabitants. As stated by a bird in Victor Delfin's painting of a tuberculosis outbreak in a shantytown of Lima: "I renounce humans. I request a bird's passport."[95] These strategies denounce the trajectory of humans, which leads to misery, in favor of other sovereignties—specifically, that of birds, who transcend the fate of humans and the state. Yet, as we will see in the next chapter, insects such as bees are the latest contested frontier in swarm intelligence at borders around the world.

FOUR

From the Eyes of the Bees *Biorobotic Border Security and the Resurgence of Bee Collectives in the Yucatán*

Imagine you're hiking through the woods near a border. Suddenly, you hear a mechanical buzzing, like a gigantic bee. Two quadcopters have spotted you and swoop in for a closer look. Antennae on both drones and on a nearby autonomous ground vehicle pick up the radio frequencies coming from the cell phone in your pocket. They send the signals to a central server, which triangulates your exact location and feeds it back to the drones. The robots close in. Cameras and other sensors on the machines recognize you as human and try to ascertain your intentions. Are you a threat? Are you illegally crossing a border? Do you have a gun? Are you engaging in acts of terrorism or organized crime? The machines send video feeds to their human operator, a border guard in an office miles away, who checks the videos and decides that you are not a risk. The border guard pushes a button, and the robots disengage and continue on their patrol.

This is not science fiction. The European Union is financing a project to develop drones piloted by artificial intelligence and designed to autonomously patrol Europe's borders.—Zach Campbell, "Swarms of Drones"

. . .

The United States competes with the European Union in the latest space race to automate the first fleet of militarized bee drones programmed to swarm and surround bodies and objects from the sky, on the ground, in bodies of water, and in outer space.[1] Swarms of robotic bees are the latest border security innovation deployed to track unauthorized border crossers around the world. This fascination with automated swarm intelligence recalls Colonel Richard Irving Dodge's and Lieutenant Arthur L. Wagner's awe of the Indian warriors who collectively swarmed the cavalry soldiers unannounced, as described in the first chapter. Appropriated again within militarized visions of border control, Apache warriors and scouts return as swarms of robotic bees at the border. Nativism has incarnated as robovision. Yet this is not the only appropriation at play here. The most numerous Indigenous migrants forced to cross the US-Mexico border are Maya from Mexico, Guatemala, and Honduras. These refugees flee their homelands, where alliances between capitalist agrobusinesses and military and paramilitary forces actively target Maya. From 1960 to 1996 over 200,000 Maya were murdered in Guatemala alone. In this chapter I ask why the United States turns to automated bees as a strategy for ubiquitous border control at a time when migration, set in motion by genocide and exacerbated by climate change, poses a significant threat, sending millions in the near future to cross national borders.[2]

Not only are Maya refugees an undertheorized constituent of Indigenous migrants across the US-Mexico border, but Maya communities in the Yucatán, mostly women, are also returning to beekeeping practices with their local Melipona bee as part of a long tradition of autonomy in the region. By turning back to beekeeping—an ancestral science that, when practiced, strengthens insurgent memory with land—Maya communities strategically band together to call their people back home and to hold on to their land in their rural jungle communities. Maya-bee intrabecoming is part of a longer story of autonomous collective selfhood steeped in the resurgence of Maya rebellion against the extractive dispossession of their lands.

Swarms of miniaturized surveillance bees extend state vision, mobility, and communication with the goal of detecting and preventing wild, rogue, or unauthorized movements across borders. Automated robotic bee drones represent another militarized technology inspired by the threat and agile maneuvers of insect-human swarms, weaponizing nature. Here again the threat of a racial other—Apache warriors, Maya refugees, hordes of Mexican migrants, and swarms of hostile bees—is co-opted and transformed from a racial threat to a natural, automated, even patriotic force in support of US empire and settler dispossession. By examining the research and development of robotic bees by the Defense Advanced Research Projects Agency (DARPA), the Los Alamos National Laboratory, and university laboratories, I offer insights into why bee swarms have become such a potent technology for border control. My research also helps explore how and why cultural beliefs and scientific studies of bees participate in contestations over land from Spanish colonialism in the Yucatán to Maya migration and US empire.

To understand how bees appropriate Indigenous knowledges—from that of the Apache warriors of the nineteenth century to that of Maya beekeepers from Mesoamerican times until today—this chapter traces debates within and across robotics and artificial intelligence, feminist new materialism, Indigenous/Chicanx decolonial theory, and evolutionary theories of bee intelligence from the nineteenth century to the present. Across these discussions, bees continue to be at the apex of debates on race, intelligence, and consciousness that are used to justify epistemic and material borders between primitive and civilized, as well as life and nonlife, or human and object. These borders matter tremendously to the politics of inclusion/exclusion, self/other, and to the managing of relation to land as property-object versus ancestor-life.

Insects figure prominently in the history of automated warfare, from the Indian Wars to World War I, when the first unmanned aerial vehicle, a torpedo nicknamed the Bug, would fly a set path, shed its wings, and release a bomb.[3] In this vein, DARPA's specialization in intelligence, surveillance, and reconnaissance has led to numerous technological "improvements" in the ability "to detect, identify, and track foes . . . and to target weapons with unprecedented accuracy."[4] And given insects' keen sensory capabilities to navigate and respond to their environments, mechanized insects such as the RoboBee, a robotic pollinating bee that I return to later, are also deployed as intelligence-gathering scouts, or what DARPA calls the "scout for insurgents," on the battlefield and at national borders.[5]

In news articles, scholarly articles, TED talks, and congressional policy reports on military and border control, swarm intelligence is characterized as nature's collective and self-governing intelligence, where decisions and actions are made without a leader dictating commands. While the previous chapter traced the unconscious drive of automated physiological movements as the colonial frontier for innovations in intelligence gathering, in this chapter bees (as well as ants, fish, and other natural entities such as cells) inspire another dark world of ubiquitous war, a surround sound of buzzing surveillance that promises an automated world of total security and control. Swarm intelligence recodes bees as scouts charged with detecting irregularities in human targets and other biological weapons hidden across a range of environments. In this chapter I trace how the government and various tech companies exuberantly adopt lessons from natural science in order to repackage insects as a *superintelligence*, rather than as mindless drones or conformist worker bees, while liberalizing this turn to nature as a collaborative and revolutionary endeavor to replace the intelligence of bees (another species on the brink of extinction) with intelligent machines, or robotic minidrones.

In order to trace theories of the swarm as a superintelligence—a hive mind—I turn back to scientific and philosophical studies of bees by Charles Darwin and others who consider human-animal evolutionary theory through bee intelligence and consciousness across Europe and Latin America. Bee intelligence sits uncomfortably at the cusp of questions about human-animal consciousness and intelligence in the context of mass industrialized labor and mass migration across borders. For instance, the status of humans' and nature's evolution changes from instinctual to superintelligent, from laborers without consciousness to intelligence gatherers. Debates among nineteenth-century evolutionary biologists were critical to the field of new materialism. Contra Darwin, German scientist Ludwig Büchner argued that bees are in fact not mindless drones who follow instinct but are intelligent creatures. I take the science of bees even further, to the Maya of the Yucatán, to offer a decolonial perspective on human-bee intelligence that dramatically breaks with evolutionary theory and the separation of humans from animals. Maya beekeeping reignites insurgent practices that counter corporate extraction and military intrusions onto Maya land. I follow the science of beekeeping from Darwin and Büchner to examine the longue durée of colonial science, steeped in extractive relations to land, and the radical sacredsciences of Maya beekeeping that are regaining traction as a life-sustaining alternative to the spread of militarized destruction.

University research guiding the study of swarms raises questions that challenge the military ethos of hierarchical control. In particular, How do these little creatures coordinate when no one individual is in charge? What drives insects not simply to work tirelessly but also to communicate decisions and to solve problems as a collective? The question of how masses of insects coordinate their collective behavior recalls colonial fascination—both in the US borderlands and among the Spanish colonialists in the Yucatán—with how Indigenous peoples like the Maya developed cohesive and autonomous agrarian societies without hierarchical structures of control and power.[6]

Feminist vs. Indigenous Perspectives on "Life"

This question regarding autonomous collectivities is one that scientists in other life sciences obsess over, such as how birds know when and where to migrate, or how bees sense the flight path that directs them home to the hive. These questions also animate philosophical concerns about the autonomy of life when no human or outside force is in charge, a question taken up by new materialist feminism and science and technology studies more broadly. For example, in Samantha Frost's recent book, *Biocultural Creatures: Toward a New Theory of the Human*, she asks a similar question about how life at the cellular level of proteins works, or how they know what to do: "Given that we do not impute intention or self-conscious deliberation to proteins, how can we account for the ingenuity, the precise coordination, and the efficiency of their actions?"[7] She argues that each act is done so that the next may be done "purposively or with a reason."[8] But she struggled with this thesis because she thought, How absurd, molecules do not have reason, and they do not intend what they do. Through such astonishment, she says, "I realized that, in spite of my theoretical training, I had a theological hangover, which is to say that I could not figure out how such processes could be possible without someone, somewhere, knowing what to do."[9] Her desire to see a single agent directing molecular life was tied, she realized, to a lingering belief in a god, or an outside force directing all life, an idea she, like many other new materialist scholars, finds to be irrational. To arrive at a more precise science entails banishing racial alterity, an archaic belief in religious, spiritual, and cosmic forces as separate from the material/earthly world.[10]

My concern is that the scientific observation of the (mysterious) life force animating collaborations among all levels of the nonhuman world continues to impose a break between a religious belief that attributes vitality to God and a materialist scientific account of life. In other words, I question the overlap between feminist materialist accounts and the military application of science, both of which rely on life's autonomy in ways that reinforce colonial racial divides between primitive unconscious forces beyond human vision and cognition, on the one hand, and civilized scientific rationality that bestows autonomy on the natural world without human or divine intervention, on the other.[11]

Maya worldviews have long understood the intrarelated ecologies of human-animal-deity worlds that continue to be banished from science and relegated to the status of primitive beliefs. As argued in this book, Indigenous sacredsciences are based on the study of animals that entail methods of becoming with these beings to extend human becoming and knowing, or what Karen Barad calls an ontoepistemology.[12] Given that many technologies today surrogate for labor that humans no longer want to do, replacing enslaved or devalued labor, many scholars refuse the binary opposition between the technological and the human and think of them along a continuum of exploitation, alongside slavery, colonization, and empire.[13] A collective of Indigenous scholars engages with AI not as an alien other but as "making kin": "Ultimately, our goal is that we, as a species, figure out how to treat these new non-human kin respectfully and reciprocally—and not as mere tools, or worse, slaves to their creators."[14] While they do not specify the ways AI imagines itself as evolutionarily replacing the intelligent force of the natural world, they refuse to treat these objects as Westerners do, as nonhuman objects void of interiority, and thus as slaves, or as machines unworthy of relation.[15] Their sense of interiority parallels that of Philippe Descola, an anthropologist of South American Indigenous cultures, who defines it as "what we generally call the mind, the soul, or consciousness: intentionality, subjectivity, reactivity, feelings, and the ability to express oneself and to dream."[16] Given that various Indigenous peoples have traditions of treating the more-than-human world as kin, their manifesto works toward imagining what it would take to incorporate AI into everyday kin networks.

DARPA's Innovations into "Life"

With its national security branding strategy, DARPA emphasizes its ability to "redefine possible" by reengineering and monitoring life from all scales. With a budget of over $3 billion a year to produce the most "innovative" strategies and technologies in national security, DARPA is one of the world's most influential and powerful military science agencies. In its mission statement (figure 4.1), DARPA brands itself with a sci-fi image of scientific breakthroughs, reinforcing the use of natural mathematical laws to uncover and journey into every dark unknown crevice of life, or future frontiers.[17] Life's (dangerous) mysteries are tamed when they come to light through a technoscientific eye that uses high-tech camera imagery to capture the beauty and mystery of invisible phenomena, from cells to outer space, all bracketed with waves of beehive patterns. By foregrounding its mission of national security through scientific research, DARPA hopes to liberalize its image from that of a secretive Cold War institution to a more transparent and democratic one.

Bee swarms are key to DARPA's motto: "Creating and Preventing Strategic Surprise."[18] Closer research on one's foe increases one's ability to launch a surprise

Figure 4.1. DARPA's innovations into "life." DARPA'S national security branding strategy emphasizes its ability to "redefine possible" by reengineering and monitoring life from all scales. Source: https://www.darpa.mil/news-events/darpa-redefining-possible.

attack and to preempt the enemy's maneuvers. Here again the shock and awe of the Apache surprise attack serves as the unconscious drive behind their approach. The agency aims to match the perceived agility of an enemy who is smaller and able to maneuver flexibly with technologies of warfare that can detect threat from all directions and scales (space, air, ground, and ocean).

To accomplish this, DARPA has switched its focus from large-scale robotic developments to smaller, more mobile, and less centralized military innovations. As stated by the director of DARPA, "Today and in the years ahead, our potential adversaries will still include nation states, but also smaller, less well defined bad actors and an increasingly networked terror threat." To get at the smaller nature of threat, he continues, "biology is nature's ultimate innovator, and any agency that hangs its hat on innovation would be foolish not to look to this master of networked complexity for inspiration and solutions."[19] The militarized innovations produced by DARPA have more recently taken a biological, neurological, and softer mechanized approach. The goal is to create swifter, more intelligent robots, while also arming soldiers' bodies with synthetically infused capabilities to make them faster and less susceptible to biochemical warfare.[20] The agency's technoscientific developments infuse warfare into all levels of society—from border security, to agricultural security (bee colony collapse), to capitalist labor efficiency and safety.

However, DARPA does not conduct its own scientific research. Instead, its team of entrepreneurial leaders, many of whom are scientists, collaborate with the most cutting-edge innovators and then hire defense contractors, academics, and other governmental organizations to materialize their creative ideas. Some of these outside research laboratories include elite institutions such as Harvard University, the Department of Homeland Security (DHS), the University of California system and the Los Alamos National Laboratory. Together, DARPA, the DHS, and Harvard researchers hope to discover what animates the mysterious movements of insects in order to mobilize this intelligence for border security, war, and the consumer marketplace. For instance, DARPA and scientists at Los Alamos have been studying bee sensing of chemical and nuclear traces since 2005 in hopes of launching bees to detect bombs, narcotics, biological threats, and migrants at various land and sea borders. In the video "Scientists Train Honeybees to Detect Explosives," they state, "Honeybees are nature's rugged robots." Further, because of bees' powerful odor sensors, "honeybees are one of nature's most sensitive detection platforms."[21] And swarms of beelike drones are being tested for a host of uses in border security, as laborers, pollinators, and much more.

The Greening of DARPA: Solving Environmental Problems

The military is one of the most egregious land-grabbing institutions that has settled on and occupied Native American land. In addition, most universities and research labs are on stolen Indigenous lands, including the Los Alamos National Laboratory, which resides on Navajo, or Diné, land. Los Alamos, similar to DARPA, markets itself with the veneer of science as a tool to solve the world's problems, ironically including addressing nuclear threats caused by bombs it helped produce, while also failing to address its own dumping of toxic materials on Native American land.[22] Institutions such as DARPA promise to securitize not only the nation but also the environment, deploying killing in the name of saving. Not so ironically, bees are used to detect the very nuclear devastation Los Alamos researchers helped create.

Perhaps unsurprisingly, DARPA's perverse turn to nature to mimic life has the goal of militarized national security. In an article from *Mother Jones*, Nick Turse takes readers down a long list of animals that DARPA mines for inspiration for more lifelike strategies of warfare and solutions to social problems.

From bees and flies to bats and geckos, DARPA is "interested in investigating biological organisms because they have evolved over many, many years to be particularly good at surviving in the environment . . . and we hope to learn from some of those strategies that Mother Nature has developed."[23] Their list includes Wolfpack: "a group (pack) of miniaturized, unattended ground sensors that are meant to work together in detecting, identifying and jamming enemy communications." Here again, the stealthy reconnaissance by Indian Scouts and the Border Patrol Shadow Wolves is automated. Similarly, DARPA is tinkering with adding living cells to robots, while the intelligence of bees is increasingly mechanized.[24] Despite the acknowledgment of nature as a sophisticated system with an intelligence that evolves with the people *and* environment, most Western technological fixes create the very environmental problems they claim to solve.

With funding from DARPA, researchers in biorobotics at Harvard University are also developing softer, smaller and smarter robotics, attempting to get at more sensual, even organic relations among nature, humans, and robots. They hope to build less mechanistic and more organic robots and to reroute AI away from the individual human brain to an organic intelligence found in nature, especially in the collective behavior of bees. In 2009 Radhika Nagpal and her team developed the controversial RoboBee Project at Harvard (with funding from Barack Obama's administration), seeking to build a robotic pollinating bee that would temporarily compensate for the loss of 50 percent of beehives from 1950 to 2006.[25] This project actually began in 1998, when researchers at University of California, Berkeley, received a $2.5-million-dollar grant from DARPA to create an insect drone.[26] This same year 2.5 million honeybee colonies were rented for pollination in the United States, many shipped to large-scale almond farms in Kern County, California. Almond growers and beekeepers quickly realized that many of their bee colonies arrived dead in commercial trucks, killed by parasites such as the varroa mite.[27] Bees were worn down by pesticides, stress, disorientation, and overwork. Clearly, there is little funding for research to address the political-military-environmental context such as large-scale industrial production, pesticide use, monocropping, bee experimentation and instrumentation, and the politics of genetically modified seeds, which all contribute to the decline in bee colonies. As argued by Jake Kosek, scientific approaches to the mystery of so-called colony collapse disorder focus on individualized biological changes in bees rather than the political-economic conditions.[28] Bee populations will likely continue to decline, and they will become mere inspiration for robotic replacements as scouts in war and as pollinators in the fields.[29] Now many teams hope to build bee swarms called Kilobots for a range of uses, including border surveillance and war, search-and-rescue missions, and the construction of buildings (rather than hives).

In addition to the proposed use of swarm intelligence to solve a range of social, environmental, military, and political challenges, the turn to nature, and to softer robotics, aims to garner public consent and to cloud our perception, portraying an inevitable world-to-come surrounded by machine intelligence. This future world will not appear as a dystopian world where clunky mechanical robots replace humans; instead, automation will fade from view as merely the next phase in the evolutionary ballet of a natural world in perfect sync/order. Robotic engineers hope machines and nature might blend into each other so that the borders are indistinct, and thus the use of machines becomes less objectionable. Drawing on a variety of nature's superspecies for novel approaches to AI, robotics is packaged as moving from centralized command and control to emergence, or from a world controlled by humans to one modeled on nature's automated collaborative and more democratic systems. Social control is dressed up as an artful dance of bees in perfect harmony.

Back at Harvard, Nagpal situates the rise of biorobotics at the intersection of computer science (AI/robotics) and biology. If robotics asks what makes things work intelligently in the world, the field of biology offers, she argues, the best example of autonomous systems. The push in AI to build robots that can act on their own behalf, without the constant direction of human operators, turns biology and nature into an ideal system to emulate, with the goal of producing robot swarms that are "self-organizing," an automated mass that uses sensory intelligence to make collective decisions. Swarm intelligence exerts a gravitational pull on all biological systems, from cellular biology to computer science, such as in the field of amorphous computing, which looked to biology and physics to structure distributed computers that network at a mass scale.[30] By turning to cellular biology, researchers analyze how cells, miniscule parts of life, work together with millions of other cells to sustain the body's life-form. And while AI developers have predominantly debated whether consciousness, as a mental faculty, could translate to robots, Nagpal and her associates want to think about computation and intelligence beyond the brain and beyond the individual. Similar to the parts of the body, which work in concert to get us up a steep hill even though we may not be conscious of each body part's role in the act, each ant, bee, or termite is imagined to be a part, or cog, that has insignificant power without the collective working together as a team. As Nagpal states during an online interview, "The brain is really the colony, no individual in the colony actually has the full picture. The full picture is distributed over these many, many minds."[31]

Automated swarm intelligence regurgitates a colonial imagery of the Indian surprise attack as recounted in Colonel Dodge's book in 1877:

> There are no ranks, no organisations or units of command; but there are words or signals of command, by which the same evolutions are repeatedly performed, seemingly more by the admirable intuition of the individual Indian than by any instruction that could possibly have been given him. The whole band will charge *en masse*, and without order, on a supposed position of the enemy. At a word it breaks or scatters like leaves before the storm. Another signal: a portion wheels, masses, and dashes on a flank, to scatter again at another signal. The plain is alive with . . . a living mass of charging, yelling terror.[32]

The evolutionary force of nature is wild and dangerous, locked into the Western psyche as a resistant force that unites a mass of Indians into an intuitive embrace with its rhythms. Nature and Indians are inseparable from a mass threat, lacking the individual creativity and imagination that allows humans to transcend or supersede nature. At the same time, Dodge remarked at the mysterious Indian invention of signaling by way of a looking glass that the chief held in his hand to communicate his commands.

The scientific study of bees by scholars and researchers is impossible to extricate from the particular historical socioeconomic norms of the day, such as industrialized production and gendered divisions of labor. Drones, or worker bees, have long served as the cautionary model for modern social control in relation to a labor regime that must, as Karl Marx suggested, strip workers of consciousness and convert them into a machine-like entity that merely follows the pace of mechanized order emerging in automated factory workplaces. Despite the elaborate hives bees build, Marx wrote, the bee is no architect, for "the architect raises his structure in imagination before he erects it in reality."[33] Western debates about consciousness begin even earlier, with René Descartes in the 1600s, who argued that animals lacked consciousness, as they merely followed a blind instinct or a mechanical impulse communicated through external organs.[34] Descartes not only viewed bees as lacking consciousness but also imagined them to be living machines or automata relegated to the status of objects, unworthy of rights or unable to make decisions. This external force directing animal behavior was interpreted by some as God, or a soul, or even an external force of energy. Others followed in this philosophical tradition of splicing apart matter and spirit, including Immanuel Kant.

A few centuries later, Charles Darwin takes up these theoretical reflections in his 1859 book *On the Origin of Species*, one of the most influential scientific studies in the field of biological evolution.[35] His chapter "Instinct" addresses, and then reroutes, the thorny question of bee intelligence, or their capacity for reason, through the idea of hereditary knowledge, or instinct. In this chapter he engages in a comparative analysis of three hierarchically stratified bees: the perfect European hive bee, the Mexican *Melipona domestica* in the middle, and the most primitive, the humble bee. Darwin proclaims the European hive bee to be one of the most intelligent bees, an assessment still revered today for the bees' impeccable creation of the perfectly shaped hexagonal hive cell—and as having the "most wonderful of all known instincts."[36] As a materialist, Darwin relies on what can be seen (honey-storage-cell structure) to explain an otherwise perplexing question: How do hive bees know how to create such perfect designs? Here he departs from eighteenth- and nineteenth-century mathematicians who saw a divine hand shaping the innate perfection of the honeybee cell, which mimicked the perfection of the anatomical cell.

Darwin departs from the science of comparative anatomy into comparative physiology, from taxonomy to behavior. For Darwin, bee behavioral knowledge was instinctual—not unchangeable impulses from God, a soul, or another outside force, but behaviors passed down through hereditary (rather than environmental) adaptations perfected through time. Archives full of documentation of ancient bees supported his argument that they had had many centuries to perfect their ways. At the same time, bee intelligence threatens to unravel his argument, as he says, "I have nothing to do with the origin of the primary mental powers, any more than I have with that of life itself. We are concerned only with the diversities of instinct and of the other mental qualities of animals within the same class."[37] Darwin brackets a discussion of the origin of bee intelligence, and theories of life more broadly (God, spirit, etc.), and focuses on his scientific comparative approach based on direct observation—a methodology of seeing steeped in the colonial context of his time—to assess behavioral intelligence *across* three races of bees. He relies on a colleague's study of South American bees to support his argument that instinct drives the evolutionary biological process of inherited descent by comparing the Mexican *Melipona*'s and the humble bee's honey-cell structures with that of the European hive bee.

In this story he argues that the perfect hexagonal shape of the hive-bee cell has evolved from that of their primitive ancestors, the *Melipona*, whose round and irregular honey-cell architecture, Darwin argued, wasted more wax and thus lacked the beauty and perfection of the hexagonal hive-bee cell.[38] Influenced by the economic culture of his time, he describes efficient and orderly production with perfection: "the motive power of the process of natural selection having been economy of wax; having succeeded best, and having transmitted by inheritance its newly acquired economical instinct to new swarms, which in their turn will have had the best chance of succeeding in the struggle for existence."[39]

For Darwin, humble bees are the oldest recorded bees, and few still exist, while the intermediate Mexican *Melipona* is destined for extinction, living on only as an ancient instinct, a foreign inner (genetic) force, passed down to the stronger European species. Divergences within a race of bee proxy for racial and national difference.[40] Otherness is not erased in this tale of biological evolution but maintained as the early origin from which the European bee species inherits its more complex intelligence, and as the pivotal force to prop up his argument about the natural progression of descent, rule, and the superiority of one species over another. His theory of descent naturalizes colonialism as the progression of history, or the temporal and spatial mapping device to trace the evolution of simple societies to increasingly complex ones, from primitive instincts to more advanced, or civilized, complexity.

So what would change if bees were seen as intelligent? Ludwig Büchner was the first scientist to argue—in his book *Mind in Animals* (1880)—that bees and other insects were in fact intelligent.[41] Büchner was a German philosopher, physiologist, and physician known for inaugurating nineteenth-century scientific materialism. Like Darwin, he was staunchly against the theological romanticism of his time that regarded God or the supernatural as the force of creativity and perfection in the world. Animals were not, he argued, a blind instrument of an *occult power*. Yet, as a defender of the exuberance of the material world and of humanity, he refused to see bees through the eyes of Darwin, that is, as mechanized adherents of instinct. Even though Büchner began to chip away at Darwin's theory of descent, he continued to use bee intelligence as a tool to banish the invisible, the unknowable, and any possibility of a force or power outside human perception. In contradistinction to Darwin, he refused categorical divides between humans and animals, providing ample evidence that bee intelligence was equal to, or even surpassed, human intelligence since bees think, observe, reflect, and take advantage of changing environmental and social conditions. His method continued to expand that of Darwin's as he studied not the evolution of comparative anatomy but that of comparative psychology, or of the mind, since the mind, he argued, could also be traced through behavioral, physiological evidence.

For Büchner, the natural world was not an unsolvable mystery but a material physical force that could be seen and known. Through the close study of bees and other insects, Büchner read the poetry, romance, wonder, and amazement of the natural world as inspiration for a better vision of humanity. He took a stab at Darwin directly when he argued for different, yet equal, intelligence across bee species. For example, the *Melipona*, he argues, produces a higher quality of honey than the European hive bee. By proving animals to be intelligent, he dispelled the theories of his predecessors; in close observation of bee behavior, he saw no divine purpose or supernatural design, no will, and no laws imposed by external forces. He was the first to think critically about the anthropomorphic blindness that occurs when human hubris and ego cloud our ability to see animals as having a similar level of intelligence to humans. In fact, he argued that bee intelligence was formed through centuries of experience in an environment, based on both long- and short-term memory, and thus was so honed that they developed extraordinary senses that allowed bees to do and see things humans could not.

In Büchner's chapter "Bee Nation," bees are idealized as an ideal monarchy where the queen reigns. At the same time, this monarchy has strong features of communism and socialism. Private property does not exist, resources are evenly distributed, and each bee eats the same thing, works collectively, and engages in selfless behavior when another is in need.[42] Most significant is their intertwined destinies—if the colony collapses, they all die together. Far from naive, Büchner sees not inherent perfection but complex situations, such as the chaos that ensues (rebellion against work, theft, and settlement of other colonies) when the queen dies or gets sick. Bee swarms merely respond to a change of power. There is a poetics to the swarm: scouts first take flight to find a new home, and then, if the colony becomes divided, swarming is necessary to reestablish harmony and order within the new hive. He brings to life their complexity to argue against the idea of an inherent and consistent hereditary instinct. Similarly, he suggests we look to bees as models for social-political transformation, and he shows the natural intelligence driving rebellion in humans and animals, critical to breaking down one order and replacing it with a better one. While Marx saw revolutionary potential in the peasant class, Büchner looks to bee collectivity to imagine a broader transformation. He wrote this during a time of political transition from the sovereign power of the divinely ordained monarchy to the nation-state. There will always be those who do not work, who deceive, and who abuse, and those who resist. This is the natural order of change that Büchner advocated.

This vision of social balance contrasted with US fears of alien invasions by hordes of migrants and other "pests" of social disorder during this same period (the late 1800s), when the media, social planners, and eugenic scientists instilled the fear that Chinese and Mexican immigrants would swarm across national borders. There were other concerns about swarm behavior during this time that speak to the anxieties of industrial production, such as the movement of newly "freed" slaves into the labor market, domestic farmers who left their rural homes for urban jobs, and international migrants from China and Mexico taking over the fieldwork abandoned by domestic farmers.

During the 1970s, media accounts warned the public about various racial varieties of swarms attacking the nation, such as images of migrants climbing border fences or a horde of Asian and Mexican/Latin American migrants invading the nation.[43] These racial threats culminated in the actual threat of a new swarm of a monstrous kind. The media tracked Africanized bee swarms traveling from Brazil into Central America and then to Mexico and the United States. Known as "killer bees" for their aggressive behavior—killing over a thousand people and numerous animals—this Frankenstein hybrid bee took flight, refusing to remain a docile tool of human design (a Brazilian scientist had crossbred the East African and numerous European honeybee varieties to produce more honey). During a time of heightened attention to racialized immigration and outsourced labor, the racialization of the swarm as "Africanized" stoked fears of social disorder, an unnatural experiment that resulted in a superspecies that crossed all types of borders (human/machine, natural/unnatural, racial, national borders, species boundaries, etc.). Seemingly driven by the gathering fury of exploited workers from around the world and armed with razor-sharp navigational directives, these bees crossed continents as if heading straight for the United States. The most well-known media event to capture this fury was the 1978 film *The Swarm*, in which a dark and violent swarm of intelligent Africanized bees kill humans who attempt to eradicate them, until the military tricks the swarm at the end of the film and kills them all.[44] Despite all odds, they continue to return.

The Invisible Hand of Automated Labor

Automated computing research has resurrected interest in swarms. At Harvard, Nagpal's team left their labs to trek to Namibia to observe and study termites. Along with other scientists, they investigated massive termite mounds, as well as bee and ant colonies, to think about decentralized computing algorithms. In an unsurprising move, nature's collective, democratic, and autonomous structures drive the revolutionary language of biorobotics. As Nagpal says, "Termites are a classic system where people have studied decentralized cooperation or implicit cooperation. The idea is that with termites, there's definitely no supervisor who's telling them what to do, there isn't even some sort of hierarchy. Instead there's sort of these chance encounters where information is propagating, but there's also just encounters with the environment."[45] In contrast to a perception of robots at the mercy of fascist humans who are in control, automation here becomes part of the natural system, which does not proceed through individual intention, nor even have a predetermined goal or plan of action. Rather than locate consciousness in the brain or in an outside (divine/sovereign) force, Nagpal's team turned to a colony of insects that feel, sense, and act consciousness through their bodies, in relation to each other and to changes in the environment. This research is clearly influenced by Rodney Brooks's AI research during the 1980s. He argued that behavior-based robots (rather than human intelligence) offer a simple, or primitive, intelligence in which behavior does not respond to a central brain that plans out actions in advance; instead, the body moves in response to the environment through nonhierarchical sensor signals.[46] Rodney's argument about human intelligence is less cited, however, than the French entomologist Pierre-Paul Grasse, who wrote about insects during the 1950s. After studying termites, Grasse coined the term *stigmergy* to characterize the process by which termites build complex structures collectively through the localized behavioral movements communicated by several neighbors moving close by.[47] Here the swarm is the location for a collective intelligence that otherwise eludes any one member.

Nagpal's team developed what they called "TERMES" in 2014, a robotic construction crew that can assemble blocks into three-dimensional structures "without any human intervention; no foreman, no central brain."[48] One day they will scale up to a thousand tiny robots, or Kilobots. In a TED talk, "Taming the Swarm: Collective Artificial Intelligence (2016)," Nagpal moves from a discussion of termites who build massive structures without talking to each other to a construction site. During her TED talk, our gaze is directed to a large screen behind her depicting silhouettes of human bodies in front of cranes taking a break at a construction site. Consistent with Harvard's promotion of a softer robotic model that will replace massive, clunky robotics, here, too, the tiny and lightweight Kilobots are cheaper and more tasteful robotic accomplices. It is clear that the "robotic construction crew" will replace precarious workers (perhaps migrant or Mohawk ironworkers) on the job site, workers who sit perched high up on dangerous high-rise buildings, who get tired and hungry and take breaks, and who labor at the bottom of a hierarchical labor structure, beneath the white-collar owners and creative/intellectual class. Not only will Kilobots end dangerous, even exploitative, labor conditions, but the thousands of tiny robots will learn how to build without the need for human intervention (and salaries). As argued by feminist and critical race scholars Neda Atanasoski and Kalindi Vora, liberal scholars perpetuate a racial logic when they argue that using robots to replace noncreative labor will eradicate human suffering and free us to do labor that is more creative (2019). Yet even the managerial or creative class is disappearing from the scene of labor.

As "low-skilled" workers are absented from the job site, thousands of fast-building Kilobots enter the scene as quietly as cells building the body, efficiently, like a force of nature itself, an unremarkable presence in human perception, even more invisible (and exploitable) than the shadowy racialized laborers who previously built other high-rise buildings. Nature—demoted to an automated, unintentional mass—here becomes ideal for manipulation and replacement, especially as these movements are imagined as occurring without the aid of human direction or intelligence. Kilobots (one thousand–robot swarms) are only strong in large numbers and designed to be individually inexpensive so they can be easily replaced.

Nagpal asks, "So the field of engineered self-assembly is basically—How do we make the molecules or the robots or the cells to self-assemble the thing we want, as opposed to the thing that they're naturally going to do?"[49] And, ironically, the swarm concept driving the Kilobots is depicted as a militarized force of nature, a docile army of insects ready to mobilize at any moment. This fascist visual iconography shares an interpretive logic with the democratic lure of bee colonies that drives researchers to reproduce bee and wasp swarms, which can be tamed and controlled, unlike the potential threat of mass human laborers, whose likelihood of revolt grows as they increase in numbers, or when they become a self-organizing mass that evolves into an intelligent, or even rebellious, swarm.

Within biorobotics, the life force of matter is segregated from humans. This alienation of material spheres preserves categorical divides among matter, animal, and human, ignoring the interconnections among the environment, insects, animals, and humans. When scientists turn to insects to revolutionize and naturalize security, thus positing biology as an automated biomass, how might the biological stand in once again as the primitive racial other stripped of human intention, agency, and intelligence, while humans disappear from the scene of racially motivated violence?

I return to the scout bees one last time. Developers at Georgia Tech study the waggle dance of the honeybee for clues to the *mystery* of bee scouting intelligence.

Using high-definition visual technologies that see in the dark, the researchers follow and map a bee colony's waggle dance. They define the waggle dance as the vibratory communication used by scout bees to communicate the precise direction and distance of a pollen source to other forager bees. Researchers believe the bee scouts communicate the precise coordinates of the sun and wind to other bees, perhaps even following natural electromagnetic forces. They may even release pheromones to indicate the amount and quality of the food source. In a marketing documentary, the researchers claim to use this information in their multiagent robotics laboratory to "help [them] uncover more secrets behind animal communication that lead to innovations in robotics."[50] What seems like a mystery and a secret to these researchers is ancient knowledge of astronomy to Maya beekeepers. Ironically, this navigational knowledge was banned, then relegated to the status of superstition, only to be resurrected centuries later as a "discovery" that will attract millions of dollars, perpetuating relations of extraction, extinction, and environmental disaster for those whose ancestral science continue to be rendered as primitive spiritual belief, in contrast to modern innovative discoveries. A similar relation to "discovery" plagues some aspects of new materialist feminist theory, such as the turn to objects and the more-than-human as vital matter. The novelty of this turn erases (and sidelines) the centuries of knowledge carefully gathered by Native peoples—knowledge that was regarded by the Spaniards, then other Euro-Americans, and now today's scientists and scholars as primitive superstition or belief.

Maya Beekeeping and the Sacredscience of Intrabecoming

Maya in the Yucatán are returning to beekeeping, a practice passed down from their ancestors, communal knowledge traced back to their ancient codices. When bee populations began to decline in the 1980s and then dropped more rapidly in the 1990s, Maya realized that there were few beekeepers left, especially those who cared for their local bee, the stingless *Melipona Beecheii*. According to Mexican biologist Rogel Villanueva-Gutiérrez, out of more than one thousand colonies known to have been maintained in 1981, only ninety survived in 2004.[51] Many elders were getting too old to care for the bees, and the younger generations were moving to urban centers like Cancún or migrating in mass numbers to the United States. Today local Maya (with the support of state and international groups) are educating each other about the necessity and benefits of *Melipona* beekeeping, resulting in a 25 percent increase in bee colonies each year. The honey and wax are becoming more marketable locally and internationally because of their curative effects for a range of conditions, including eye diseases, gastrointestinal illnesses, infertility, and asthma.[52] This is a significant comeback for bees who were on the verge of extinction in 2005, with their populations having decreased by 93 percent over the previous twenty-five years.[53] And the revival of beekeeping is tied to the hope that fewer children will have to migrate to urban areas of Mexico, or even to the United States, and that migrants will return home.

What Darwin considered to be the natural evolution of the stronger honey bee species over the *Melipona* covered up centuries of colonialization, which continues today. Maya understood the loss of local bees as a symptom of centuries of extraction from their land, which caused forests and jungles to disappear. They remember when this happened before, first with the Aztecs who commanded large quantities of tribute of wax and honey. Then, during the eighteenth and nineteenth centuries, Spanish colonists introduced monocropping in the Yucatán, which led to deforestation and another decline in bees and production cycles.[54] When forests are torn down, the plants that only the *Melipona* pollinate are removed, along with their nesting homes in the trees.[55] In addition, plantations replaced the traditional Maya *milpas* and slash-and-burn agricultural methods.[56] Despite this decline, Spanish rulers increased demands for tribute of honey and wax. Given this extra stress on bee production, the respectful relationality between the bees and the keepers was greatly diminished, leading to lower honey production rates. To keep up with the high demand for honey and wax—the two most traded commodities before and after the Spanish conquest—the Spaniards imported the *Apis mellifera*, or honeybee, to the Americas. Importing the honeybee led to competition for floral resources between the *Melipona* and the less discriminating honeybee, and soon the higher-producing honeybee replaced the local bee population. When this happened, traditional beekeeping diminished, accompanied by an attempt to destroy the knowledge of a people, their origin stories, and the intimate practices that served as rituals of relationality between the Maya and land. Even today government-funded apicultural workshops, education, and knowledge sharing on European honeybee breeding across Guatemala threaten traditional *Melipona* beekeeping. The pride of bringing back traditional beekeeping is bound up with centuries of fierce social movements by Maya, who have long fought against attempts by the Spaniards and others to dispossess them of their land, extract their labor, and eradicate their knowledge practices.

According to contemporary studies in the Mexican Yucatán, women continue to make up 40–75 percent of the caretakers of the bees, many of whom are also *curanderas* (healers) who tend to women's fertility cycles and crop production, using the honey collected for their healing practices and for trade. Many new recruits become politicized through their participation in beekeeping collectives. One of the many female collectives founded by Maya women, Koolel-Kab (women who work with bees)/Muuchkambal is an organic farming and agroforestry organization that fights for forest conservation, promoting Maya land rights and environmental education and programming to reverse the effects of industrialization and deforestation. They have shared their organic beekeeping model with more than twenty communities as an alternative to illegal logging. After constant agitation spearheaded by Leydy Pech (a Maya woman, beekeeper, and activist), they won a legal battle against Monsanto in 2014 to ensure that Maya communities are consulted before any large-scale agricultural projects can be approved.[57] Protecting the jungles of the Yucatán is key to their identity as Maya (rather than as Ladino or Mexican) and continues the centuries of Maya perseverance as autonomous despite the many waves of colonial persecution by the Maya elite, Spaniards, white mestizo Mexicans, the Mexican military, and multinational corporations.[58] Given the dense jungle on the eastern and central Yucatán Peninsula, Maya were able to sustain their practices and land, whereas more populated and less forested regions were easier for the Spaniards to surveil and control.

When the Spanish did attempt to encroach on their land or demanded overly burdensome tribute, Maya communities could easily flee into the jungles, temporarily or more permanently, to evade capture. And they have a long history of revolt against being incorporated into colonial rule, assimilated, and dispossessed of their land. Many still consider themselves independent from Mexico, both long before and after the Caste War (that ended in 1901), although they claimed themselves an independent republic from 1841 to 1848. They refuse to give up their independent status, thus posing a threat to the Mexican government and developers, who fear another revolt may grow. It may not be readily apparent that insurgency grows through women beekeeping collectives, which preserve a strong sense of ancestral identity with the land and rekindle a past that helps them envision and enact autonomy as Maya into the future.

Their motto speaks to their activist commitment across a range of issues such as the protection of forests, waterways, bees, and women: "Koolel-Kab: Polinizando un mundo justo y sostenible" (Pollinating a just and sustainable world). With a vibrant online presence, they discuss the double risk of being a woman and an environmentalist and thus refuse to disentangle the fight to protect their land from respect for women's contributions to and value in the home and community. This grassroots collective has posted events and talks online by Maya and Latin American Indigenous activists such as Leydy Pech and Berta Cáceres, a Honduran environmental activist, Indigenous leader, midwife, and cofounder of the Council of Popular and Indigenous Organizations of Honduras. Cáceres fought for land rights, especially against dams, illegal logging, and plantation land grabs in Indigenous communities. She gained local acclaim for her grassroots campaign that led the world's largest dam builder to pull out of the Agua Zarca Dam at the sacred Gualcarque River.[59] This acclaim, however, led to a spate of militarized violence targeting dozens of women environmental activists. Cáceres was killed in her home by a national militia trained by US forces. The protests over her death and those of countless other female environmentalists sparked a worldwide movement in support of their grassroots efforts. Violence at all levels affects women, from corporate degradation of their land to violence against women in the home and by the military. Yet they continue to put themselves on the line to protect what sustains them, their communities, and the world. And they are not alone. Thousands of Indigenous peoples around the world who fight to protect their forested land are secretly murdered by "hit men, security guards, private contractors, ranchers, and timber gangs" supported by corporations, bankers, and military regimes in Latin America and trained by the United States' School of the Americas.[60]

Despite another severe decline in the local population of bees (similar to global patterns around the world), Maya today are returning to ancestral bee rituals, calling back the bee spirits that will wake up the land through reciprocal practices of care and love. The *Melipona* bees are the only ones who pollinate the diverse tropical plants in the forests of the Yucatán and Quintana Roo region, and thus they decolonize this area by fertilizing the tropical plant varieties that sustain the diverse human-animal-ecologies that have been plundered by tourism and development projects that have resulted in massive logging of Maya rural land. The return to beekeeping has also meant a return to ceremonies and rituals lost when other bee species, such as the honeybee, dominated.

The Erotics of Ancestral Beekeeping

A series of precisely detailed scientific engravings of the *Melipona* bee and detailed instructions on Maya beekeeping were recorded in twelve extant pages of the Madrid Codex (also known as the Tro-Cortesianus Codex). According to the Mexican historian Laura Sotelo, if you compare the practices of beekeeping in the codices and traditional beekeeping of the *Melipona* today by the Maya, you will find "an amazing parallel."[61] For the ancient Maya, the science of beekeeping reflected the erotic relationality between bee and beekeeper. In

Figure 4.2. Beekeeper from the Madrid Codex, 1250–1450 AD. Image courtesy of the Foundation for the Advancement of Mesoamerican Studies.

the drawing in figure 4.2, the bee shaman can be seen with three fingers touching the bee's mouth, with his belt/member visible. In response, the *Melipona* is drawn with its eyes crossed in ecstasy, expressing a similar labor that ignites feelings and becomings of cosmic union for both bee and beekeeper. Pleasurable bonds of care are forged between bees and Maya in the creation of practices that build mutual existence and plentitude for all.

By gesturing with three fingers, the beekeeper communicates many layers of meaning held by the sacred number three. The bee is often depicted as part human, part deity, and part animal. The shaman also honors caring for the bee as a beneficial ritual that unites and fertilizes each of the three sacred realms: the sky, earth, and underworld. Submitting to the spirit, or gravitational pull, of the earth becomes a sensual labor that communes across these realms.

In the Codex; on ceremonial temples; and on pottery in Tulum and Coba, Mexico; the *Melipona*, or stingless Yucatán bee, is often depicted as a goddess (Ah-Muzen-Cab, or "royal lady") with her wings and legs spread open to the sky and head pointed to the ground as if landing while in flight (see figure 4.3). With its crossed eyes, the bee in flight is split between worlds, a bee in ecstasy—taking in nectar and touching the earth—while the bee shaman who touches the nest/hive travels with the bee across the spiritual and material realms, from the cosmos to the underworld (figure 4.4). While invisible to the human

Figure 4.3. Ah-Muzen-Cab (the descending god), Temple of the Descending God in Tulum, Mexico. Source: https://growinggreener.blogspot.com/2012/11/various-gods-of-beekeeping.html.

eye, the sun and moon pull and push the plants' flowering, as the moon pushes the tides of water under the earth and the heat of the sun pulls the plant out of the earth.

These erotic ties to land as animate were dangerous to Spanish hierarchical religious orders that demonized human-animal-deity existence. How could Spanish friars command labor, devotion, and obedience to their hierarchical world order (with God at the top) so alien/ated from the pleasurable (and fertile) bonds with land found in Maya religious texts and practices?

I first began to rethink human consciousness many years back through Gloria Anzaldúa's embodied practices of *chamania*, or the female shamanism she refers to as *la naguala*. In Anzaldúa's approach to studying precolonial, Aztec, and Mesoamerican female deities, shamanism, and communication with animal spirits through dreams, visions, and her own childhood memories, I found an attempt to expand border consciousness and ontology.[62] She resignifies and rematerializes border consciousness from human-centric concepts of racial mixture (*la mestiza*) to human and more-than-human crossings (*la naguala*). While both terms engage with questions of Indigeneity, Anzaldúa imagines beyond prohibited borders confining nationalist conceptions of subjectivity,

Figure 4.4. Ah-Muzen-Cab (the descending god). Temple of the Descending God in Tulum, Mexico. Image courtesy of the Fundación Melipona Maya.

sexuality, and race and also extends consciousness across sensual becomings and relationality with land.

At the same time, important critiques of Anzaldúa, for instance, articulate some of the problematic ways la mestiza draws on Indigenous cosmologies from the past to forge a Chicanx borderland subjectivity "to the exclusion, and indeed, erasure of contemporary Indigenous subjectivity and practices on both sides of the border."[63] This appropriation, argues Domino Renee Perez, allows Anzaldúa to individualize Indigenous traditions into a grab bag of mixed tricks, bypassing important questions such as how Anzaldúa's disruption of categories to invent new ones resonates for specific Indigenous peoples.[64]

It is worth examining some of Anzaldúa's less analyzed, unpublished, and newer work that takes seriously the imperative to cross Western borders dividing the human from the nonhuman, as a violence that continues to delegitimate Indigenous knowledge systems. I especially appreciate Anzaldúa's poetic musings on how human and more-than-human entanglements erotically transform us, reshaping not simply our mind but our very material form, thereby refusing US categorical borders segregating knowledge and land. Anzaldúa's desire to expand consciousness, or knowing, into the dark (places unknowable by Western, Mexican, or even human intelligence) involves an embodied form of visioning based not simply on scientific observation but on chamania, on shape-shifting into the spirit of the other—often animals—to extend human knowledge and vision. Thus, consciousness (*conocimiento*) grows the more we enter into other realities, the more we enter the darkness, or knowledge worlds deemed outside the human and the real.[65] Whereas Western science relies on *technologies* to extend human vision under the pretense of objectivity, for Anzaldúa, each animal spirit offers itself as a technology, with its own particular vantage point and aptitudes for how it sees, senses, travels, and knows. Thus, to know otherwise involves an alternative science where we change our form and become other.

In her unpublished writing, Anzaldúa taps into a deeply embodied sensual connection with the cosmos as part of the process of writing:

> And then I put a root down into the earth through my cunt and connected to the middle of the earth. I shot out another tentacle out of my head and into the sky to connect with the heavens. Then I just concentrated. And somewhere here, about a foot above my head or in front of my forehead I locked in. You know how two magnets lock in? And I'd bring it, the consciousness, up. And with my eyes slightly crossed, I focus there. . . . After that, it happened again a few more times. A connection so incredible. In sync—my heart, the heart of my cunt, Her heart, the heart of the Cosmos, beating all in one rhythm.[66]

This connection to the forces from above and below is critical to her sense of the spiritual, or the sacred force that drives all life, a binding force felt in the body, a force that holds us together in place. According to anthropologist and astronomer Anthony Aveni, Albert Einstein, like some other scientists, "spoke of a 'cosmic religious feeling,' that lay behind some of his eureka moments concerning the harmony of nature."[67] These moments that others have expressed as surges of insight deeply felt in the body also speak to moments when "we sense that all boundaries between ourselves and the outer world vanish, for what we are experiencing lies beyond all categories and all attempts to be captured in logical thought."[68]

This moment of cosmic unity depicted in the Maya codices and described by Anzaldua helps us rethink the creative processes of writing, beekeeping, and scientific inquiry more broadly, moving away from Western notions of labor and toward a practice, or act of creation, tied to worldings, rituals of creation that inspire new intimacies with worlds, or other ontoepistemologies. Anzaldúa's chamania can be thought of as a labor not geared toward the production of an object but instead a sacred enacting of worldings that defies Western ways of knowing that presume and enact separation, difference, and subject-object relations. And although Anzaldúa has been understandably accused of extracting Mesoamerican Indigenous knowledge in ways that relegate Indigeneity to the past, I contextualize her embodied practices with animal spirits in conversation with Indigenous studies, Western science, and contemporary Maya beekeeping cosmoscience to raise some points of contact but also important differences between Maya and Chicanx cosmologies. Her cosmic visionings and becomings with other forces transcend the here and now of human time/space. I was reminded of this moment of ecstatic connection in Anzaldúa's writings when I came across the images of the Maya beekeeper/shaman.

Anzaldúa's writing cracks open moments when she slows down to enter into communality with the spirit worlds she intuits in small but elongated moments and queer crossings with nonhuman time, space, and beings. To commune with is to acknowledge the life force of the earth and cosmos, a cosmic knowledge practice that is important to recognize as its own science rather than a belief system. In a similar vein, Patrisia Gonzales discusses Mesoamerican knowledges as a cosmologic, where "indigenous peoples know the cosmos as alive and interact with it as a communicative being that transmits knowledge to humans and all sentient beings."[69] This practice of communing-with that shapes Anzaldúa's feminist and queer methodology in *Borderlands/La Frontera: The New Mestiza* is a model in this book for transcending epistemological, ontological, and species borders; in her writing she becomes la naguala, able to transform herself into a tree, a coyote, another person. Anzaldúa engages erotic relationality beyond the human and is queer in her promiscuous affinity for women, goddesses, gay men, and the not-so-human, whether this means folks who are pushed out of the category of the human (the queer, the squint-eyed, the perverse) or the sensual forces all around us that the West regards as passive and unchanging matter, as raw material for human intervention and profit.

In her unpublished writing, Anzaldúa tells a story of la Prieta, a young dark-skinned girl who has an innocent adventure with another young girl in the fields, a playful encounter that wakes up a sensuality all around her (Anzaldúa says things like, the air cracks with desire and "the roots of the tree uncurl and stretch out tautly").[70] These erotic feelings of being gravitationally tied to a place and one's surroundings and of falling into erotic rapture inspired by the more-than-human are inseparable. For Anzaldúa, these deep meditations that directed energy from the center of her body—heart/cunt—up to the sky turn her writing from labor to ritual, from the making of a product to the creative act of fertilizing new worlds across the borderlands in the midst of displacement and threats to render her relational ties to land extinct.

As a "de-tribalized" and "de-Indianized" Xicana, Anzaldúa's life's work has been to dig up Aztec and Mesoamerican sacred knowledges and practices that were stolen through Spanish colonization and US empire.[71] Her relation to Indigeneity is fraught with nostalgia and invented traditions, even as there is something powerful in the possibility of building a political affinity with Indigenous peoples today through the way she holds on to her homeland-turned-borderland through an intimate relationality with land that follows the footprints of Aztec/Maya ancestral presence. While for Anzaldúa these long-studied skills allow her to disrupt Western notions of land as a primitive female body to be conquered, contained, and exploited, I turn to Maya practices of beekeeping to offer a bridge between decolonial Chicanx/Latinx erotic worldmaking and the radical practice of beekeeping by mostly Maya women in the Yucatán, both of which are invested in reclaiming relational belonging with land. How might Anzaldúa's decolonial sensuality take on a more critical charge when situated in solidarity with Maya beekeepers' sensual relation with bees in the context of their own displacement across borders, and also alongside a long radical tradition of fighting to hold on to their land and autonomy?

Queer scholar Mark Rifkin, who does not claim Indigeneity, finds inspiration in the poetry of Cherokee, Two-Spirit poet Qwo-Li Driskill in his book on the "erotics of sovereignty," or the feelings of connectedness to the land that enact sensual belonging that defies the bureaucratic state apparatus that defines identity through law (personhood), property relations, and governance. For Rifkin, land is not an inert object but "sensate and desiring . . . a mutuality of becoming."[72] An erotic orientation to the land as erotic lover also shifts the focus from sexuality (and reproduction) as a human affair to a broader co-evolving dynamic that unites consciousness, obligations, and collective transformation with the nonhuman. The time of the world moves not along a progressive path from point A to point B, as if abiding by its own sovereign tempo, but in relation to the entangled desires of humans and nonhumans that cause speedups and slowdowns in consciousness.

Kim Tallbear, an Indigenous science and technology scholar, also aims to decolonize settler sex by rethinking the possibilities of polyamorous sexuality within an Indigenous context of relation rather than regarding it as merely an act or identity.[73] Tallbear leans on Indigenous scholar David Delgado Shorter's research with the Yoeme peoples who live across the US-Mexico borderlands, especially with the *moreakamem* (healers or seers among the Yoeme), who taught him to see sexuality and spirituality as inseparable powerful forces efficaciously used toward healing.[74] Tallbear says that given that the moreakamem "have reciprocity and receive power in their encounters with spirits, ancestors, dreams, animals and humans, sex and spirit become sets of relations—through which power is acquired and exchanged in reciprocal fashion among persons, not all of them human."[75] The values of interconnectedness across all aspects of life that are shared by many Indigenous peoples are described by Shorter as "intersubjective," or "mutual connectivity, shared responsibility, and interdependent well-being."[76] Shorter sees great potential to extend our intimate relations into broader kin networks as we reimagine our sexuality to look "more like a type of power, particularly one capable of healing."[77]

For many Maya collectives, the labor of beekeeping and writing is pleasurable, a political practice tied to their ability to be autonomous, to enact a complex worlding that enables them to regenerate a cosmology of living and healing rather than to be relegated to the margins, destined for extinction. How might these practices of caretaking nurture reciprocal relations with the spirited presence around us and materialize a return of the people, animals, and plants to the land? Once this deep attachment, desire, respect, and attraction dies out, the gravitational pull of interconnection between Maya and bees will fall away.

These erotic and material enactments go against the Western imagination of the bee's behavioral "nature," associated with docility or threat, that dominates the imaginary of robotic bees as ideal tireless pollinators and as invisible eyes of surveillance deployed at the US-Mexico border. Instead, Maya have long interpreted the labor of foraging as sacred, related to the bees' role in origin stories as the creators of life. *Melipona* bees were connected to fertility, the transporters of life from one world to the other. The honey was offered to women to boost fertility and soothe labor pains, and it was used to sweeten a drink called *balaché* used during Maya ceremonies of regeneration that called forth and gave thanks to the continued motion of all realms of life. Nestled inside a tree trunk, or *cajon* (see figure 4.5), you'll find an inner world that looks as much like a cellular structure as it does planets connected in the universe. Maya consider the *Melipona* nests as a sacred womb, especially as the tree (*jobón*) is known to connect the underworld to the sky, from the roots to the leaves. *Melipona* represent and enact a womb or portal translating between timespace worldings, connecting the earth with the divine.

One of the stingless bees native to the Yucatán, or the Xunán Kab (The Royal Lady/*Melipona beecheii*), is known to have predated humans, birthing a world of plant life crucial for human flourishing. The bee is regarded as a sacred being, one of the few animals domesticated by precontact Maya, and the honey and wax are revered for their healing properties and considered divine as the bees travel to the celestial cosmos (*Xmahen*) to consume the honey, which is then transported down to the land to fertilize the earth.

Figure 4.5. *Melipona* beehive. Photo by author.

The Mesoamerican science of beekeeping is a sensual practice of intrapenetration across the three levels of the cosmos (the underworld, the earth, and the sky, or astrological realm). The bee was depicted in the Madrid Codex with the numerical value of three, as bees were often figured in images together with the jaguar and deer, or as a trio of human, insect, and goddesses. The jaguar was known by the Maya as having supernatural and transformative powers (crossing worlds), and the deer was the protector of the hunters, the bees, *and* the beekeepers, who were often priest-shamans.[78]

These bees translate/transport visions and worldings from the cosmic to the earthly realm. On the top of the temple at Coba is where the Xunan Kab is engraved, a sacred place to study astronomy, have visions, and engage in purification rituals and ceremony.[79] The ancient science of astronomy, tied to the Maya calendar, numerology, agriculture, navigation, and origin stories, was banned by the Catholic Church as the stories competed with the explanatory power of the Spaniards' god. Origin stories and knowledge of astronomy were replaced with biblical stories of Jesus and the dominance of anthropocentrism and the biological sciences.[80] Much of the astrological knowledges recorded in the codices were burned in 1562 by the Spanish missionary Fray Diego de Landa as they were thought to be superstitions that distracted the Maya from carrying out their duty to God. Despite this massive destruction, these knowledges endure today through agricultural practices, artistic patterns woven into fabrics, the celestial rhythm of agricultural planting and harvest, architectural designs, and beekeeping practices.

For Maya today, human-insect-spirit worlds evolve together through a vibrational consciousness and an erotic exchange of knowledge and becomings. In fact, some stingless bees, such as the genus *Melipona*, vibrate their bodies to extract pollen.[81] The migratory path of bees, sucking nectar from one flower and then another, is a sensual journey of mutual flourishing across species. In this ritual practice and labor tied to fertility and food production, pleasure, productivity, and spiritual connection are inseparable. Without the *Melipona* the Maya jungle will perish and disappear, along with the bees and people. Thus, the bees and people enact sacred codes of life through their "labor."

The beekeepers take from the bees in the most respectful and reverent manner possible. Even today it is common to hear Maya beekeepers refer to these bees as people, and many speak to the bees with reverence and care. Don Cristiano, a shaman-beekeeper with whom I spoke in a small village thirty miles from Coba, in the Yucatán region of Mexico, said he visits and speaks with his bees often, promising them he will never leave them. And they are loyal in return. He explained that "although they work hard all day, they never travel too far, and they always return home." He then paused and said, "Like us." Even when hurricanes hit, he added with pride, they manage to find their way home. Out of Don Cristiano's eight children, all but a few had migrated to urban areas such as Cancún to work in tourism.[82] With the declining prices for agricultural goods and the economic restructuring from agriculture to tourism, Don Cristiano and his wife also depend on tourism to survive. Some of the ecotour outfits include a visit to his humble home, where he discusses the importance of beekeeping, while his wife and youngest daughter make beautiful hand-sewn *huipiles*, or blouses, to sell to tourists like myself.

It is common for Maya to refer to the *Melipona* bees as similar to them. Thus, like the Maya, they are revered as being gentle (yet aggressive when necessary) and sensitive to the emotional vibrations of anyone they come into contact with. Don Cristiano explained that they know him by the vibrations he gives off, and thus he tries to rid himself of any fears, problems, or worries when he speaks with them. He placed a bee on my arm, and I was amazed by their small

Figure 4.6. *Melipona* bee on my finger. Photo courtesy of Neda Atanasoski.

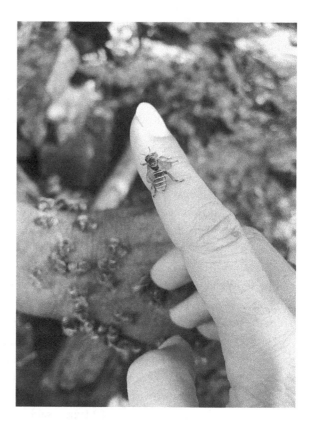

size yet strong vibrational force, tickling my arms and fingers as they left tiny imprints of pollen (see figure 4.6). I understood the erotics of communing as it became clear that this little being, who looked directly at me, was speaking to me, demanding my attention with vibrational force as I kept a close watch, laughing in utter amazement and wonder.[83]

Through close observation, ceremony, dialogue, and ritual, the stingless bees are a model, or perhaps mirror, for the Maya people's own nonhierarchical society, where bees, women, and the earth were equivalent terrain for fostering the continuation of life for all.[84] Maya observation of and intimate closeness with the bees in ancient times as well as in the present provide important lessons for respecting and understanding the cycles of life and death, growth and reproduction, and especially how to avoid the overexploitation of natural resources. Like Maya who caution against overharvesting, the bees will not grow the colony unless there are sufficient resources owing to favorable environmental conditions.[85] Maya turned to the bees as a revered technology, as an "ecological indicator to prognosticate the future success or failure of the food production in agricultural trends."[86]

Shared Destiny between Maya and Melipona

Abejita Melipona . . .
que hermosa centinela
la que cuida tu panal
la que da su vida entera
por la reina y el lugar
eres tú mi bella abeja
la que endulza mi existir
eres tú la melipona
la que me hace sonreír
no destruyan más el campo
no acaben mi existir
dejen que abunden las flores
para que yo pueda vivir
gracias

(Little Melipona bee . . .
how lovely is the guardian
who cares for your honeycomb
the one who gives their entire life
for the queen and the place
it is you my lovely bee
the one who sweetens my existence
it is you the Melipona
who makes me smile
don't continue to destroy the fields!
don't end my existence!
let the flowers grow abundantly
so I can live.
thank you.)
—"Poesía 'Abejita Melipona'"

• • •

Maya see their destiny as the same as the bees' destiny.[87] In the video "Xunan Kab" the camera shifts from a scientist discussing beekeeping to a young girl reciting in her Mayan language a love poem, manifesto, and prayer spoken as much to the bees, her people, as to those who cut down the jungle where she lives. She declares her love for the bees as inseparable from the love for her people and their land, as their destinies bind them together. The Mayan language does not segregate spheres of life, or separate the world into separate genres, species, or categories of thought. The honeycomb is also the home or land of the Maya, and the reciprocal care of self-as-other blends into a practice, worldview, and material intrabecoming with the animals, land, and people. In "Xunan Kab," we learn the ancestral sacredscience of beekeeping through numerous perspectives and fields of study.

Women beekeeping collectives mobilize land-based activist movements alongside the protection of the bees and their people. A short clip by the women of Koolel-Kab called "Soy Abeja Maya" ("I Am the Maya Bee") tells a simple but powerful story that builds an ontoepistemology of Maya sacredscience.[88] The statement, demand, and provocation "I Am the Maya Bee" serves as a methodology, ontology, activist praxis, and approach to a sacredscience of interbeing that renders unthinkable Western Cartesian divides that sever our bodies and integrated ways of thinking and being. I am you; you are me. She speaks through a worldview embedded in the Mayan language that many Xicanx and Latin American Indigenous women/feminists understand as a praxis of interbeing.[89] This concept is embedded in the Maya concept "In'Laketch," or "you are my other self," to imagine beyond the duality of Cartesian rationality that divides self from other, which is the root of an autonomous, self-reliant subjectivity upholding human exceptionalism.[90] M. Bianet Castellanos describes another form of personhood from the Maya Tzotzil (people of the bat), who are "taught that self-awareness is rooted in a collective identity."[91] This radical notion of the self as less important than the collective is being taken up across a range of AI uses in robotics and beyond. The idea that the swarm collective is more intelligent than the individual is a Maya concept that caused Indigenous peoples to be regarded as primitive in evolutionary stories and then today is being stolen as the basis for Western innovations in the science of robotics. Across numerous configurations of swarm intelligence, the collective is touted as smarter than the individual.[92]

In Carlos Lenkersdorf's study of the Mayan Indigenous language, Tojolabal, he understands their language through intersubjectivity, where the concept "we" is at the core of their language.[93] Universal notions of culture (The Science or The Culture) do not exist, nor do notions of objectivity and subjectivity. His students tell him, "Brother Carlos, you know our communities. If there's a problem, we don't go each one to his home to resolve it alone. On the contrary, we get together to resolve the problem because, look, here we are 25 students. You tell us, who thinks better? One head or 25 heads? And here, each one of us has two eyes that makes 50 for all of us. Tell us again: which one has a clearer vision: two eyes or fifty?"[94] This ancient knowledge that continues through Maya worldviews is revered when "discovered" by scientists in nature but regarded as primitive and even criminal when this knowledge goes against the imperatives of capitalist extraction and the profit of the few over the well-being of the many.

Conclusion

Technological control over nature, such as the development of Robobees on the border, pretends to be a form of automation that is naturally self-regulating. Not only do these technologies continue to be driven by the fear of nature and Indigenous people's mysterious synergy with the natural world, but swarm surveillance also (mis)appropriates Maya beekeeping science, passed down by generations and documented in the codices. Maya beekeeping offers an appropriate technology for going beyond Western science that alienates the world into objects. Western computational theories of evolution driving automated bees erase the violence committed against Maya lifeworlds when land and knowledge are stolen, such as the intimate ties of becoming across humans, bees, deities, animals, the sky, and the earth. It is no surprise that these knowledges were often passed down by women, who staunchly continue to defend their ancestral sacredsciences with bees and collectively labor to rebirth the return of their land, people, and animals.

Part of the book's goal has been to uproot the evolutionary drive justifying technological innovations in militarized border surveillance tethered to settler historical concepts such as civilization and progress. To do this I have foregrounded the military's arsenal of "natural" intelligence through the use of Indigenous, animal (RoboBees), and plant (saguaros, as we will see in the conclusion) technologies inspiring automated surveillance operations premised on the need to replace worlds that will inevitably go extinct. Border technologies see through secular binaries such as life/death, self/other, secular/sacred, human/nature, and so forth. These divisions prevent settlers from understanding the sacredscientific relational knowledge practices with the more-than-human that are misappropriated for use in securitizing the border, including surveillance technologies—from Native eyes, to drone vision, to automated sensors, bees, and saguaros. Rather than replace the natural world with technological fixes that presume the coming extinction of Native peoples, the bees, and even the land, Indigenous peoples such as the O'odham, Maya, and Apache carry on ancestral knowledges to call their land and people back. The current migration of Maya is predicated on a colonial plantation system that similarly dispossessed Indigenous peoples of their land and then extracted materials that were critical to their flourishing with land. The Maya ancestral sacredscience of beekeeping is a practice that triggers memories of resurrection that hold coded knowledges that will once again free Maya communities and the bees from colonial and neocolonial dependency and dispossession.

Beekeeping is an integral ritual that is inseparable from everyday life. This ritual is sacred because it strengthens the freedom of bee-human-deity to hold worlds together across various scales that support a wealth of life-forms. By tracing the movement of the bees, Maya follow in the footsteps of their ancestors, remembering knowledges triggered in their limbs, mind, and heart, a kinetic knowledge or ritual of movement activated through gestures, prayer, intentional words and movements. Thus, the practice of caring for the bees imprints the origin stories of human and bee becoming, of their descent from the universe to earth, into their DNA. If the forests disappear, the bees fade away, then the people, too, will disappear as the memory of what makes them who they are dies out. The stronger the bond between the *Melipona* bee and the Maya people, the more fertile the land will grow, building harmony across all forms of life when they are able to autonomously live, honoring the distinct, but united, calling or path they are meant to follow.

Conclusion *Wild versus Sacred: The Ongoing*
Border War against Indigenous Peoples

The proposed Wall is a desecration of spiritual and sacred Indigenous Land.
—Nellie Jo David, Hia-Ced O'odham activist and scholar

• • •

The Indian Wars endure across the US-Mexico Border, on O'odham sacred land. On this land, border occupation and dispossession recently spiked to the level of unspeakable devastation and violence. Sidestepping consultation with O'odham tribal leadership, in December 2019 President Donald Trump authorized the construction of the first section of a thirty-foot-high steel wall on a region of the Arizona-Mexico border under federal protection called Organ Pipe Cactus National Monument.[1] Adjacent to the official boundaries of the O'odham reservation, this expansive borderland holds O'odham ancestral burial sites, artifacts, and rocks marked with petroglyphs by their ancestors, the Hohokam, who lived there between 200 and 1,400 CE. Many O'odham experience border construction to be a war zone. Bulldozers and explosions destroyed O'odham burial sites and gutted sacred land, where ancient saguaro cacti were dismembered and tossed to the side of the road.[2] O'odham tribal members place their bodies across the border in protest, locking arms and standing in front of construction tractors to halt this destruction.

This massacre is a direct attack on the O'odham, who witness the destruction of their past, present, and future. To destroy these sacred sites is to sever their knowledge practices and intimate ties with, and protection of, their ancestral land. In preparation for the erection of the wall, contractors blasted burial sites where the bodies of O'odham *and* Apache ancestors had been laid to rest. Congressman Raul M. Grijalva explained, "Monument Hill is the resting place for primarily Apache warriors that had been involved in battle with the O'odham, and then the O'odham people, in a respectful way, laid them to rest on Monument Hill.[3]

Also scattered across this land are the bones of over 2,700 Mexican, Central American, and Maya migrants—unburied, unsettled, and mostly unclaimed. While Apache, O'odham, and Maya have distinct histories and stories, their paths cross in this contested border region. Our journey ends here, on this land where their footprints touch and overlap from many directions, times, and places, entangling their land-based struggles and demands for sovereignty against a dictatorial US military power that controls the border and well beyond.

In this final chapter I end with the border on Organ Pipe to highlight the violent ways federal conservation and national security converge on "wild" land and contribute to the dispossession of O'odham and criminalization of migrants. Toward the end of the chapter I hone in on O'odham sacredscience with the saguaro cacti to reiterate my argument across the book: Native American knowledge practices reveal a deep intrarelational science that has long protected land that is rightfully theirs to steward. Chapter 2 described the effects of dispossession on O'odham tribal members via the infrastructural occupation of US border security. O'odham describe Trump's massive border-wall project—including the sprawling network of surveillance towers, sensors, and drones—as beyond words, an "unspeakable" violence against them as a people. To describe this violence as unspeakable is to understand that it is impossible for the O'odham to translate for the federal government the longue durée of colonial trauma caused by the dispossession, and now infrastructural occupation, of their land. To explain sacred relationality in the English language enacts colonial violence by framing the world with words that prop up technologies and laws that de-animate Indigenous peoples' land along the border, making it into wilderness, a void, an empty and savage space in need of security, development, protection, and conservation. But perhaps most significant are the unknown effects these ancestral disturbances will have on present and future generations of O'odham.[4]

The border wall consolidates centuries of long-term military occupation, containment, and warfare that violently rips open the land, heart, and spirit of the O'odham, who persevere against all odds to carry out their sovereign, or sacred, responsibility to care for their ancestors, land, and each other. As articulated powerfully by O'odham tribal member Ofelia Rivas, "I am an original person of these lands. I have long strived to be a voice for the plants and animals, the water and mountains, and honored O'odham that have survived since the beginning of time to continue our honored sacred obligations to live here."[5] Against the language of rights, recognition, and citizenship documents, Rivas reminds readers that O'odham ancestors have cared for this land since long before the settlers arrived, an understanding of belonging possible only through an intimate lived relation with sacred land.

The federal government designated 40 percent of Native ancestral land across the two-thousand-mile US-Mexico borderland as "wilderness" for preservation and conservation. This "public land" includes national parks, wildlife refuges, wilderness regions, historic sites, public lands, and critical security infrastructure, including water-delivery systems. In fact, if you combine Sonora, Mexico, with public land on the Arizona side of the border, this swath of the borderlands constitutes the largest region with the most protected area in North America.[6] To cordon off these sites as wild or historic, however, is to greenwash the ongoing violence of settler colonialism in conservation projects that have cleared the land of O'odham, and then of Mexican residents and migrants, to render the land *wild* in the first place.[7] The Department of the Interior (DOI) manages the majority of federal conservation land, including the 515 square miles of Organ Pipe Cactus National Monument first stolen from the O'odham in 1853 when the United States purchased southern Arizona and southwestern New Mexico through the Gadsden Purchase.[8] The national mission to protect wild land dovetails with the goal of national security, extending settler surveillance as a project that aims to protect the nation's borders from Native American and Mexican threats. To successfully build his wall, Trump would have needed not only a tremendous budget but a tremendous amount of land, creating what could have been one of the nation's largest civil and military land-acquisition projects.[9]

The incommensurate relation between O'odham and US management of land was evident in a 2020 hearing between O'odham tribal leaders, DOI officials, and Organ Pipe board members titled *Destroying Sacred Sites and Erasing Tribal Culture: The Trump Administration's Construction of the Border Wall*.[10] During the hearing, members of the DOI refused to honor O'odham demands to immediately halt the construction of the wall. In particular, O'odham challenged the department's sovereignty through appeals to international laws that mandate consultation with tribes based on their nation-to-nation status, using these laws to elevate the acts committed against the O'odham to the level of a war crime. Despite numerous attempts to remind officials of O'odham status as a sovereign nation, government representatives reinterpreted their role through the benevolent albeit disingenuous lens of liberal protection: "Under President Trump's leadership, the federal government is not only tackling the national security and humanitarian crisis, but also addressing the environmental crisis impacting the character of the lands and resources under the federal government's care."[11]

The conflation between national security and environmental protection is not merely a matter of twisted semantics. Even in the nineteenth century, long before its designation as a national monument, Organ Pipe was of interest to the federal government, which established a US customs and immigration station at the contiguous land of Sonoita and Lukeville in what is now Arizona. However, this land was not withdrawn from the public domain until a 1907 Presidential Proclamation (35 Stat. 2136), when President Theodore Roosevelt declared land within sixty feet of the border as government land for use by customs personnel.[12] Years later, the federal government took more land at Organ Pipe in Arizona. In a 1931 memorandum, a park naturalist at Grand Canyon National Park, argued that this land possessed "sufficient scientific and natural interest to qualify for national monument status."[13] There are surely benefits to the federal government preserving and protecting land. However, such claims and the resulting landgrabs have a long settler history of trumping the relational claims to sacred stewardship by Indigenous peoples.

In 1937 President Franklin D. Roosevelt finally designated Organ Pipe "a pristine desert wilderness" through authority granted to him by the 1906 Antiquities Act, passed by his fifth cousin, President Teddy Roosevelt. The Antiquities Act aimed to protect Indigenous burial sites from grave robbers, given Native bones were profitable archaeological objects of "national" interest. After 1907, however, Native bones became the legal property of the federal government and thus collecting bones by military officials, scientists, and archaeologists was legitimated without approval by Native Americans for the good of the national patrimony, university research, and museum curation.[14] It was commonplace for military officials to dig up the bodies of Apache right on the battlefield, driven by studies of craniology that aimed to prove Indians were early inhabitants of these lands in the past tense, destined to become extinct.[15] As valuable spoils of war hung on the walls of the Army Medical Museum and later the Smithsonian Institution, Apache bones materialized archaeological proof of a primitive past, bodies archived, labeled, and displayed to violently educate the public about Indigenous peoples' inevitable supersession by a more advanced race.[16] National parks such as Organ Pipe Cactus National Monument claim to serve not only landscapes, plants, and animals but also to protect "the safety of our citizens" and to preserve "some of our Nation's unique natural and cultural features."[17] However, through the guise of legal "protection," the management of grave sites was transferred from Native American to federal hands. Human remains were transformed from relational ancestors into archaeological objects; and sacred ways of seeing, and interacting with, land were replaced by settler ways.

As happened during the development of other US national parks during the early 1900s, dispossession of O'odham land took place under the auspices of preservation. During the 1940s and 1950s, officials from Organ Pipe Cactus National Monument cooperated with federal agencies to erect a border fence along the park's thirty-mile territorial border with Mexico. The wall's purpose was to control the intrusion of cattle and livestock onto Organ Pipe, as well as to prevent unwanted "foreigners" from crossing into the United States after World War II. To this end, the monument border was monitored by horse patrols.[18] Much later, this region became an important illicit crossing point for immigrants (as discussed in chapter 2) after Operation Gatekeeper funneled migrants into more desolate areas of the border.

In a related sleight of hand that erases the militarized violence and genocide that border control has committed against immigrants and Indigenous peoples, the US government justifies its occupation of O'odham land in the name of protecting the O'odham *and* their land from immigrants who supposedly threaten to turn the land "wild," or return it to a savage and lawless frontier. As stated by Paul A. Gosar, a DOI officer:

> "Uncontrolled, illegal immigration is an overwhelmingly destructive activity. You can see through these photos we're showing you [*he shows a slide of trash left behind in the desert, as the absent presence of border crossers*]. It is an activity that has deeply scarred the border regions from the San Diego Wildlife refuge to Organ Pipe National Park to the Rio Grande at Big Bend National Park. Drug running, human trafficking, trash, feces, water pollution, damage to . . . springs and seeps, foot and illegal vehicle transit . . . has left deep scars in environmental destruction across the landscape. If the Federal government were required to consider the environmental impacts on this open border policies, under NEPA [National Environmental Policy Act], the preferred alternative, we would definitely need a border wall!"[19]

Gosar's deployment of the wall as humanitarian, protecting O'odham and other wilderness land from invading migrants whose presence commits environmental violence ("deep scars" on the land), erases the scars caused by development and infrastructure, as well as the more deadly violence of the state's laws, policies, and technologies, which result in the deaths of thousands of migrants and destroy diverse habitats in this desert region.[20] His use of migrants as a pretext for the state's own role in destroying Native land ecologies and historic sites also conveniently forgets how the government has historically removed environmental and labor protections and has made the border a lawless zone for *maquila* (factory) labor exploitation and a dumping ground for toxic factory waste, especially on the Mexican side of the border. And border walls in Arizona exacerbate flooding and soil erosion and block critical passageways for numerous animals, including jaguars, bighorn sheep, and other species. As I've made apparent in the book, the militarization of Indigenous knowledge with the goal of occupying and dispossessing their land has all but erased the long-preserved practices of Native stewardship, which the Apache, O'odham and Maya continue to engage as they care for the land. Against the historical narrative of border security are the many ways that Customs and Border Protection, the Department of Homeland Security, and the Department of the Interior contribute to the insecurity of many, including funneling migrants into deadly desert regions. As shown in chapter 2 the construction of walls, surveillance towers, and buildings by the government crisscrosses O'odham land in violent and destructive ways.

Trump's thirty-foot steel wall intends to project a visual technology, or monument, declaring national victory over savage, or wild, invasions across the borderland. As such, the wall figures as a sign of dictatorial power that attempts to gag and suppress dialogue, consultation, legislative constraints, or responsibility for the extreme harm and violence it enacts. To realize his border-wall project, Trump's administration exploited a clause in the 2005 Real ID Act to waive over thirty laws created to protect the land, including the Native American Graves Protection and Repatriation Act, the Environmental Protection Act, the Endangered Species Act, and the Archaeological Resources Protection Act.[21] Bypassing these laws permitted the administration to blast another sacred burial ground next to Quitobaquito Springs and even to siphon water at the border for mixing concrete for the wall, permanently destroying the only natural spring in this desert region, which has attracted people and animals since the beginning of time. The government attempts to own and control not only the land but resources deep underground, siphoning water from beneath the O'odham nation.[22]

Unfortunately, many groups who have publicly come out against Trump's border wall contribute to the unseeing of the ways border security technologies enact violence against Native Americans living across the border. Even some of the most vociferous environmental groups that have protested against the border wall since the 1990s advocate for border *reform*, such as a more environmentally friendly wall that would allow animals to migrate across.[23] In addition, Democrats, such as President Joe Biden, join the call to stop Trump's construction of the border wall with an unfortunate plan to amplify a technological approach to the border through the construction of even more virtual walls, or surveillance towers.[24] More significantly, national parks and monuments, like Organ Pipe, under the management of the federal government, converge in their role to protect the land and nation from threats, such as "illegal" migrants, cartels, and drug traffickers, cast as "invasive species" along the border.[25] Since 2002 national park rangers have been tasked with protecting the landmass in their stewardship by doubling as Border Patrol agents. This charge, along with the hiring of extra park ranges, came after the 2002 death of a twenty-nine-year-old park ranger in Organ Pipe while pursuing a drug cartel from Mexico. Extra park rangers were hired to fortify the border. Between 2002 and 2010, an average of ten rangers "arrested 71 people, apprehended 1,231 illegal aliens, and intercepted 7,563 pounds of marijuana. . . . NPS [National Park Service] staff have [also] recovered the remains of 184 individuals."[26] As if these varied skills (including policing, military drug control, and even forensics) were not enough, a 2011 NPS document titled "Wild Matters" states that the NPS supports "national security initiatives in and around Organ Pipe Cactus National Monument, including: joint border enforcement operations; the construction of a vehicle barrier and pedestrian fence, and the establishment of a network of high-tech surveillance towers."[27] Trump transferred an undisclosed number of NPS rangers from around the country to work alongside Border Patrol on the border while his proposed 2020 budget cut $587 million (17 percent) from the entire national park service's budget.[28]

Even Organ Park's scientific approach to environmental stewardship supports settler surveillance, especially by monitoring the wilderness terrain along the border for aberrant signs of harm or danger. With the crackdown on immigrants and an increase in Mexican agriculture affecting the park during the 1980s, scientists began a monitoring program "to understand the condition of the ecosystem to better protect it from growing outside threats."[29] In 1994 this monitoring program was dubbed "the Ecological Monitoring Program" to highlight UNESCO's approach to climate change: to promote "integrated monitoring."[30] As stated in a document on the Organ Pipe website, "The methodologies and tools for long-term monitoring will assist park managers with tracking the 'vital signs' of the monument ecosystem."[31] Here we see how the science of conservation relies on automated surveillance to protect "life," a bounded territory that must remain free from outside intervention or harm in order to thrive.[32] It becomes clear why Indigenous sacredsciences are often overlooked in scientific responses to climate change based on protecting land from human interaction, as if nature thrives only if left to its own evolutionary rhythms. Ironically, it is the government, settlers, and developers who threaten to turn the land into a barren wilderness. In addition, the extra security and paperwork—as well as rules that forbid fires—have discouraged O'odham from entering their traditional land on Organ Pipe, as well as engaging in ceremonial relations with each other and the land. Over time, the land may become wild if uncared for by O'odham relations.

Descriptions of "wild" land emptied of Native Americans' footprints during the nineteenth century perpetuated the ideology of settlement, dispossession, and conquest by European Americans who traveled west.[33] As we saw in chapter 1, military innovations in surveillance, reconnaissance, and communication dependent on Native eyes extended the army's ability to control what they saw as a wild, savage, and dangerous frontier crossed by Indian and Mexican insurgents. At the same time, nineteenth-century conservationists such as Teddy Roosevelt advocated for saving "pristine landscapes" while traveling west, contributing, as Jean O'Brien argues, to their role as discoverers of an empty land, while violently erasing the footprints on the land of the Indigenous peoples who had carefully tended to these regions for centuries.[34] Despite Roosevelt's eugenic views of protecting wild land in order to secure white masculine virility, those against the destruction of his statue today at the American Museum of Natural History in New York attempt to separate Roosevelt's racist views from his status as a "pioneering conservationist."[35] Within this settler colonial context emerged the 1964 Wilderness Act's definition of wilderness (for purposes of conservation and protection of nature) as land unimpaired by human interference, or "an area where the earth . . . [is] untrampled by man, where man himself is a visitor who does not remain . . . an area undeveloped . . . retaining its primeval character and influence, without permanent improvements or human habitation."[36] This notion of wild not only reinforces the need for constant monitoring of land to prevent human crossing but also encodes the necessary emptying of land of harmful elements, including immigrants and the Tohono O'odham. It fails to recognize the centuries of footprints across these lands, the centuries of learning with the land, of learning how to care for the sacredsciences that the O'odham have cultivated and passed on, generation after generation, from time immemorial.

For many Indigenous peoples, *wild* pejoratively denotes land left untended and uncared for.[37] Wild land dies off without proper respect and attention. The costs of the state's attack on the O'odham through "physical and psychological war tactics" are high, including "diminishing women's knowledges and men's responsibilities, and . . . future generations."[38] Given women's vital role in passing on sacred knowledges, interruptions to the practicing of this knowledge not only enact another form of violence against women, but enact violence on the land itself.

Sacred Ground

At the 2020 Congressional hearing, *Destroying Sacred Sites and Erasing Tribal Culture*, New Mexico Congresswoman Debra Haaland, a member of the Laguna Pueblo tribe (who was recently sworn in as the first Native American secretary of the DOI), spoke powerfully in defense of O'odham sacred burial grounds:

> But a sacred site that's been blasted, it can never be made whole again. I want you to understand that. And you know why? Because ancestors put those things in the ground with care, and love and tradition *and prayers*. Those can never be regained again. . . . They put things there because they knew in the future we would rely on that knowledge and knowing that those ancestors are there. I don't expect you to understand that, but I'm trying to impart a little information on you so you know why they cry when they see those places being blasted apart. The damage that this administration is doing to this area is irreparable. And you didn't even ask. You didn't ask, nobody asked these people if you could do that. It's shameful. We're having this hearing because we care deeply about what's happening on this land. But it's happening all over. And I say again, I don't know how any of you sleep at night.[39]

Haaland's anger resonated across the room, targeting the hubris of federal officials who fail to consider what these burial grounds mean to the O'odham. Haaland rejects the authority of the federal government to preserve and conserve land along the border by articulating the long-lasting violence and trauma they commit against the O'odham but also against their own psyches ("I don't know how any of you sleep at night"). What was not said at this meeting was that the O'odham tribal leaders had agreed to the installation of surveillance towers on their land as a compromise; in return, the federal government had assured them it would not construct a fortified border wall, which almost all O'odham oppose. Consistent with many years of broken treaties, Trump trampled these agreements to build his wall.

O'odham land is defined not by maps and borders but by footprints marking the path of the ancestors from the past to the future, including plant and animal ancestors that have shaped O'odham existence, culture, and character as a desert people.[40] To clear the land for Trump's wall, contractors from the US Army Corps of Engineers and Customs and Border Protection uprooted many ancient saguaro cacti, sliced them into pieces, and left them strewn across the desert floor. For the O'odham, this amounted to a barbaric massacre of their ancestors that left their bodies amputated on the ground (see figure C.1). Some officials tried to appease the O'odham nation with unsuccessful attempts to replant intact saguaros in other areas such as Tucson, but most of the cacti died. These saguaros only survive in the delicate eco-system where they were born. With deep time behind their ways, O'odham watch their own potential displacement as they are uprooted and displaced onto smaller plots of land.

Without knowledge of O'odham ways, it is difficult to understand the myriad ways the deep roots attached to a particular place have great import, meaning, and future-oriented instructions. Saguaros are not plants but people/ancestors, understood as such based on their shared roots to this land and long relationships to feeding successive generations of people materially and spiritually long after those individuals are gone. Like the ancestral burial sites, the bodies of the O'odham and saguaro are buried together on this land and they rise up from the land with the spirit and nutrients of each other pressed into their DNA. One cannot simply uproot the delicate ecology and spiritual ties these ancestors nurture with the land (the saguaro thrive only in this particular desert region). The lessons the saguaro provide, the animals that sustain them, and the people who walk past and remember these many relationships are rooted in this land.

Figure C.1. Amputated saguaro cactus. Source: https://www.latimes.com/world-nation/story/2020-02-26/border-wall-saguaro-cactus.

Scholars have mobilized Indigenous understandings of more-than-human personhood to demand that rights extend from humans to mountains, rivers, and so forth. But to the O'odham, to know the cactus as a relative is to live with it intimately through ancestral practices that alter one's becoming with the other, an understanding that one's very being, flesh, and life in the desert are inextricably tied to and with the saguaro. O'odham rely on each part of the cactus plant—harvesting syrup from the flowers, seeds to make flour, and the dead ribs for fences and furniture. Saguaros provides food, shelter, and protection to hundreds of other species in life and death. Upon decomposition, its rich flesh provides necessary nutrients to the desert soil.

Like an elder, an ancient saguaro shares countless lessons on how to thrive (living longer than humans) in a desert climate. Living for up to two hundred years and weighing over a ton, saguaros rely on two root systems, one that shoots straight down to five feet and another that sprawls like rhizomes close to the surface. Through centuries of experience, saguaros know that sometimes they will draw only from surface rain, in competition with other shallow-rooted plants, while other times, when water is more plentiful, such as after the monsoon season, the deeper roots will draw from the water table farther below. And they know how to conserve this precious water. Their pulpy flesh can store enough water to outlast long droughts and their waxy skin prevents moisture loss in the arid environment. When the desert becomes barren, climate change doesn't account for the many years non-Native farmers, mining companies, and the state diverted the Tohono O'odham nation's water table to highly populated cities like Tucson. As we saw in chapter 2, just as the state does not respect O'odham sovereignty over airspace, so too the state has seized the minerals and water beneath their soil. Once the O'odham farmed the land, made fertile with water from the aquifer. Now, as a result of the state's siphoning of water (which the O'odham consider theft), the land has become nonarable.

O'odham follow the rhythms of the cacti that they understand to live in pluridimensional time. O'odham associate the saguaro's blooming period with key seasonal changes across all life in the desert. One can look up into the sky to watch the clouds gather and know this marks the end of the harvest season. To prevent the fruit from fermenting, it must finish growing and be picked before the monsoons arrive. Women follow in the footsteps of those who came before, teaching the young ones how to harvest the fruit, used to make syrup and wine, both of which are tied to the seasonal and ceremonial time of the O'odham calendar. The ripening fruit is also a call to prayer and ceremony, setting the time for elders to cross a great distance with medicine bundles.[41] The ritual momentum of these movements across borders is compromised by the Border Patrol, who vigorously search elders' sacred objects for hidden drugs. The border destroys the ceremonial communing between O'odham and saguaros, a form of sacred-science that not only sets the tempo of O'odham movements but also gives the community the ability to assess the past and future health of the land, such as by assessing the abundance or extinction of bats and other pollinators that are critical to the food chain of many others. Flat-nose bats who feast on the cacti's fragrant night flowers give back by aiding the cacti's survival, dropping seeds in the ground where a select few will take root and grow. Given the slow growth of saguaros, they grow spikes as defense against threats to their survival. Also multipurposed, spines tell stories by holding torn pieces of migrant clothing, witness to life passing by and sometimes falling to the ground. Sacred places are rooted land where life and death rise and fall in continual motion, holding pain and trauma but also serving as a model of how to battle against the forces of adversity and thrive.

My focus on the Indigenous borderlands, where Apache, O'odham, and Maya footprints collide, tells three different but intersecting stories. They tell stories of Indigenous peoples who continue the long struggle against their own erasure by fighting to defend their land. Some, such as the Chiricahua Apache, fought with the Cavalry, Mexicans, and even some O'odham before being violently removed from their land and confined to reservations. Now the O'odham resist further attempts to take their land, especially under conditions of "protection" under militarized occupation and conservation. And when the O'odham fight to protect their sacred sites, they protect Apache burial grounds and migrant remains. The Maya, too, call back the bees and their people under threat of extinction by tourism, monocropping, and global consumption of exotic wood and products, which drive corporate extraction backed by a military force with genocidal aims. By following in the footprints of their ancestors, they dig roots in and with the land that call back Maya refugees who have been displaced and scattered, many forced to cross this very same desert borderland.

Even the bloody Apache wars did not wipe out the Apache, who return to Fort Huachuca seasonally to remember and honor the many who battled to protect their land. They return with ceremony to heal their severed relation with a homeland that still calls for their return. Through harvesting, songs, music, and prayer, they awaken the Huachuca Mountains, land, and ancestors with their care, massaging the land that holds their footprints until they return.

Anti-Indigeneity characterizes US approaches to the US-Mexico border and is shared by other settler colonial borders globally. These presences and knowledges exceed settler colonial sovereignty (founded on managing land through the eyes of scientific expertise, private property, and borders and security). Moreover, these footprints constitute a threat when they block settler management of more swaths of land. The borderlands emerge as yet another rugged terrain fought over through military surveillance technologies that ironically are based on Indigenous sacredscientific intrarelationality with the more-than-human world.

I have reoriented the optic of militarized surveillance from the tracking of immigrant threat in order to also see the understudied absent presences of Indigenous occupation, settler colonial dispossession, and ecocide. Deeply informed by Gloria Anzaldúa's radical writing on the border in the late 1980s, I have returned to what she called *la herida abierta* (the open wound). While she argues that Indigeneity remains a resource for healing and living otherwise today, her writing was inspired by Aztec and Mesoamerican footprints from the past, while not addressing the O'odham, Apache, Maya, and so many more Indigenous peoples that continue to live and fight against the colonial violence of border surveillance and security.[42] The war on the border continues the project of US settler colonialism and the empire of borders around the world.[43] Border-security management is an arm of US empire, spreading the frontiers of US territory and militarized control and occupation.

Thus, rather than see Indigeneity as a past that haunts the border today, I have brought the past and present of Indigenous knowledges to the forefront of a struggle for control over territory, and of a struggle in the name of protecting national security and the environment. I have argued that Indigeneity is not merely code programmed into the ways these border technologies see and unsee. These technologies also contain and eradicate sacred relation with land through a security regime of terror that secularizes life into alienated relations that attempt to incarcerate land and people as property for both profit and control through the colonial lens of extinction. It is my hope that Indigenous/Indígena knowledges that unite them with their sacred land are seen as scientific methodology that is taken seriously in their claims to land.

As I have shown throughout the book, it is the settlers who remain foreign on the land since the time they stepped onto Turtle Island. The science and technologies used to secure their settler occupation of Indigenous land are fragile in comparison to Mother Earth. For instance, on January 29, 2020, a stunning video captured a broad section of Trump's recently installed thirty-foot-high panels careening to the ground, as strong winds literally blew the border wall down. Many prophecies, such as Anzaldúa's, have seen an end to US technological empire and the return of the land to Native peoples. In her 1987 book *Borderlands/La Frontera: The New Mestiza*, Anzaldúa writes:

The sea cannot be fenced,
el mar does not stop at borders.
To show the white man what she thought of his
arrogance,
Yemayá blew that wire fence down.
This land was Mexican once, Was Indian always
And is.
And will be again.[44]

Yemayá blew that fence down. Wire, metal, and virtual fences will fall, rust, and be dewired. As the empire spreads, so, too, do the Native peoples and allies in a broad-based protest that includes the earth's forceful ancestors. The waters will flow, the animals will cross, and the flowers will bloom again so the land and people will heal and regenerate the gift of life for all.

PREFACE

1   See María Josefina Saldaña-Portillo's *Indian Given: Racial Geographies across Mexico and the United States*, especially chapter 3, for a close reading of how the Treaty of Guadalupe Hidalgo differentiated "savage tribes" or "tribus salvajes" from "good" Indians and Mexicans, those considered "gente de razón"or "people of reason" (139).

2   According to Anzaldúa, Ana Castillo, and other Xicana feminists during the 1980s and 1990s, Indigenous knowledges and lineages had been stolen from Chicana-mestiza-Mexic Amerindians by patriarchal rule, such as by the Aztecs of Tenochtitlán (as when Moctezuma massacred thousands who dreamed about the decline of their empire, thus ending the practice of visions and dreams through a reign of terror against the spirit world). During the colonial period, the Spanish imposition of a racial caste system based on pure blood (*limpieza de sangre*) morphed into the nationalist Mexican elite (who invented the Mexican race by disappearing Indigenous peoples over time through racial mixing of Spaniards and *indios*). See Castillo, *Massacre of the Dreamers*. For a discussion of Spanish colonial race relations in Mexico, see Martínez, *Genealogical Fictions*.

3   I am using the term *Chicano* here deliberately to mark its masculinist usage. Otherwise, I use *Xicana* to specify a mixed heritage that includes Indígena ancestry or *Chicanx/Latinx* to refer to a gender-inclusive terminology for those of us living in the United States.

4   This genealogy includes a wide array of scholars who see the decolonial as a perceptual tool for making other worlds possible. These include Mexic-Amerindian-Xicana writer Ana Castillo (author of *Massacre of the Dreamers: Essays on Xicanisma*) and Chicanx queer historian Emma Pérez (author of *The Decolonial Imaginary: Writing Chicanas into History*). Other works include Muñoz, *Cruising Utopia*; Gómez-Barris, *The Extractive Zone*; Ahmed, *Queer Phenomenology*; Alexander, *Pedagogies of Crossing*; Sandoval, *Methodology of the Oppressed*; and Lugones, "Toward a Decolonial Feminism." For those who consider spirituality as a way of seeing beyond Western analysis, see L. Pérez, *Chicana Art*; and the work of the early decolonial Martinique scholar Frantz Fanon, *White Skin, Black Masks*.

5   These terms are used by Patrisia Gonzales in *Red Medicine* and extrapolated by Susy Zepeda in "Queer Xicana Indigenous Cultural Production." Detribalization signals

the historical and intellectual traditions that alienate Xicanas from their Indigenous lineages, spiritual practices, and land-based connection and culture due to the ongoing effects of colonization (i.e., forced migration and removal from land for resource extraction) and the logics of racialization that aim to eliminate "the Indian/el indio."

INTRODUCTION. TRACKING FOOTPRINTS

1 I am borrowing from Mishuana Goeman's use of *unsettling* in her keyword discussion of land. See Goeman, "Land as Life."

2 Miguel Díaz-Barriga and Margaret Dorsey argue that in contrast to Wendy Brown's (*Walled States, Waning Sovereignty*) and Peter Andreas's (*Border Games*) assertion that the militarization of borders speaks to a decline of nation-state power in the face of transnational flows of goods and people, they instead see "a reconstitution of sovereignty based on practices associated with necropower and states of exception." See Díaz-Barriga and Dorsey, "Introduction," *Fencing In Democracy*, 8–9.

3 On the absenting of Native peoples, see Jean O'Brien's book *Firsting and Lasting: Writing Indians out of Existence in New England*. There are some exceptions to this widespread erasure of Native peoples, such as María Josefina Saldaña-Portillo's *Indian Given: Racial Geographies across Mexico and the United States*, Nicole Guidotti-Hernández's *Unspeakable Violence: Remapping U.S. and Mexican National Imaginaries*, and Christina Leza's recent book *Divided Peoples: Policy, Activism, and Indigenous Identities on the U.S.-Mexico Border*.

4 Ganster, *U.S.-Mexican Border Today*; Leza, *Divided Peoples*.

5 Andrea Smith, Jodi Byrd, and Joanne Barker all discuss the complexity of Indigenous absence that is always a presence, even if as a specter of the past. A. Smith, *Conquest*; Byrd, *Transit of Empire*; Barker, "Territory as Analytic."

6 An emerging group of scholars is fiercely remapping the US-Mexico border as the Indigenous borderlands in their dissertation projects. See Salomón Johnson, "Returning to Yuma"; Madrigal, "Immigration/Migration and Settler Colonialism"; Painter, *Bordering the Nation*.

7 Attempts to naturalize current migration as foundational to the origin story of the United States render invisible American Indians' long presence on this land, their current struggles against ongoing genocide, and the forced movement of African slaves, displaced Maya refugees, and Latin American migrants. See Blackwell, Lopez, and Urrieta, introduction.

8 Saldaña-Portillo, *Indian Given*, 6; Byrd, *Transit of Empire*; and Dunbar-Ortiz, *Indigenous People's History*. Also see P. Deloria, *Playing Indian*.

9 Blackwell, Lopez, and Urrieta, introduction; Castellanos, Gutiérrez Nájera, and Aldama, *Comparative Indigeneities*; Gómez-Barris, *Extractive Zone*; A. Simpson, *Mohawk Interruptus*; Alberto, *Mexican American Indigeneities*. Also see Kelly Lytle Hernandez's book, *City of Inmates*, for an important revision of the carceral history of Los Angeles that begins with the attempted elimination of Native peoples.

10 As I discuss in chapter 1, Mexico agreed to sell southern Arizona to the United States on the condition that it would control Apache who raided Mexican villages in what was to become the northern Mexico border with the United States.

11 The Indian Wars took place from 1866 to 1891 although Native Americans across Turtle Island feel like the war against them is ongoing.

12 I use the broader term *Indian Wars* to refer to the more than thirty wars the United States (and Mexico) waged with Indigenous peoples during the nineteenth century.

13 For an excellent discussion of how Indigeneity ontologically grounds US settler colonialism, see Byrd, *Transit of Empire*. For the ways security discourses are deployed in relation to the threat of Indigeneity, see Moreton-Robinson, "Writing Off Indigenous Sovereignty," 89, 95. And for the ways Indigeneity founds the imaginary of the US-Mexico border, see Saldaña-Portillo, *Indian Given*.

14 Byrd, *Transit of Empire*. Within discussions of the rise of border control, even those that predate the federal hiring of the Border Patrol in 1924, reference is made to the Texas Rangers and the mounted guard, yet there is no discussion of this earlier period of the Indian Wars and of Indian scouts in particular. See Hernandez, *Migra!*

15 P. Deloria, *Playing Indian*.

16 Kevin Bruyneel contends that the year 1871, when the United States ended all treaty relations with Indigenous nations, marks a significant shift in US-Indigenous relations. Bruyneel argues that the post–American Civil War era, when the United States expanded its reach across the Indigenous lands of the western frontier, was a key moment that paved the way toward US nationhood and assertions of state sovereignty, which relied on "drawing indigenous communities further within U.S. boundaries without fully integrating them," thereby imposing modern colonial rule on Indigenous nations and tribes. For more on his conception of a "third space of sovereignty" in Indigenous claims to land, see Bruyneel, *Third Space of Sovereignty*, 65.

17 I discuss the Shadow Wolves in more detail in chapter 2, but their fame as manhunters is referred to explicitly in an Amazon series, *Shadow Wolves: ICE's Native American Manhunters* (Topics Media Group, 2009).

18 Hidalgo, *Trail of Footprints*.

19 Indigenous studies scholars have long debated political contestations over claims to Native American tribal membership. For some of these debates, see Kauanui, *Hawaiian Blood*; Barker, *Native Acts*; Tallbear, *Native American DNA*. For a historical discussion of how Spanish notions of pure blood (*limpieza de sangre*) influenced racial classifications in colonial Mexico, and especially the replacement, or erasure, of Indigenous claims with a mestizo identity, see Martínez, *Genealogical Fictions*.

20 Ahmed, *Queer Phenomenology*, 16. Ahmed's queer phenomenological approach to material orientation in space draws from postcolonial understandings of race and Judith Butler's queer theory of performance in the making of selves and worlds but does not acknowledge the importance of orientation in Indigenous philosophy. See Butler, *Bodies that Matter*.

21 In her theorization of the alliances between postcolonial and decolonial theory, Aimee Carrillo Rowe describes Xicana (mestiza) identity, Indigeneity, and land as incommensurate given the ways Xicana decolonial claims to Indigeneity oftentimes reinforce what she calls a "Settler Xicana" relation to the same land that Native Americans reside in and claim as their homeland. See Rowe, "Settler Xicana," 525. Susy Zepeda also engages a queer Xicana Indígena methodology to move away from

the settler logic of Chicano claims to Aztlán that fail to acknowledge other tribes and peoples on the same Southwest territory. She uses the term *Indígena* to refer to Indigenous peoples in Mexico, who are distinct from Native peoples in Turtle Island, or the United States. See Zepeda, "Queer Xicana Indigenous Cultural Production," 126.

22 See Anzaldúa, *Borderlands/La Frontera*. I am also thinking here of Eve Tuck and K. Wayne Yang's article "Decolonization Is Not a Metaphor." My critical reread-ing of Anzaldúa's becoming with the more-than-human aligns with the material-spiritual activism in Elisa Facio and Irene Lara's *Fleshing the Spirit: Spirituality and Activism in Chicana, Latina, and Indigenous Women's Lives*. Also see Anzaldúa, *Light in the Dark*. And while Anzaldúa's reclaiming of a Mesoamerican Indigenous cosmol-ogy has been widely critiqued for situating Indigenous people in the past and has contributed to the erasure of Native Americans living in the borderlands today, I engage her less acknowledged sensual or erotic relations with the more-than-human alongside current practices by the O'odham (chapter 2 and conclusion) and Maya (chapter 4). For more on the critical engagement of Anzaldúa's work in rela-tion to feminist new materialist engagements with the nonhuman, see Schaeffer, "Spirit Matters" and Kelli D. Zaytoun, "'Now Let Us Shift' the Subject."

23 Byrd, *Transit of Empire*.

24 Benton and Straumann, "Acquiring Empire by Law," 2.

25 la paperson, *Third University*, 6. The relationship between humans and the control over nature goes back to the book of *Genesis* in the bible.

26 la paperson, *Third University*, 6. Here la paperson quotes from Nelson Maldonado-Torres, who says, "'I think, therefore I am,' is actually an articulation of 'I conquer, therefore I am.'" See Maldonado-Torres, "On the Coloniality of Being," 252.

27 You can find this discussion in the chapter "Instinct" in Charles Darwin's *On the Origin of Species*.

28 Morgan, *Ancient Society*, 95.

29 Shaw, "Marx and Morgan."

30 This is an ad for Alexa, the female-voiced speaker who responds to human com-mands. See "PDR Marketing—Ad from 2021-02-27," *St. Louis Post-Dispatch*, accessed March 13, 2021, https://www.stltoday.com/ads/other/pdr-marketing—ad-from-2021 -07-09/pdfdisplayad_17650e10-1e45-5f88-9f60-d4a304ca8d55.html.

31 I am borrowing the term *surveillance capitalism* from Shoshana Zuboff's book *The Age of Surveillance Capitalism: The Fight for a Human Future at the New Frontier of Power*.

32 Hochman, *Savage Preservation*, xii.

33 Jean O'Brien argues that the "vanishing Indian" was not simply part of written nar-ration of the nation but also found in pageants, commemorations, and monuments. See O'Brien, "Vanishing Indians."

34 Hochman, *Savage Preservation*, xv.

35 Hu, *Prehistory of the Cloud*.

36 Yandell, *Telegraphies*, 5.

37 Hu, *Prehistory of the Cloud*. Also see Marez, *Farm Worker Futurism*; Nakamura, "Indig-enous Circuits."

38 Atanasoski and Vora, *Surrogate Humanity.*

39 Haraway, *Staying with the Trouble.*

40 Lewis et al., "Making Kin with the Machines." The authors quote Vine Deloria Jr., who wonders "why Western peoples believe they are so clever. Any damn fool can treat a living thing as if it were a machine and establish conditions under which it is required to perform certain functions—all that is required is a sufficient application of brute force. The result of brute force is slavery." See V. Deloria, *Spirit and Reason,* 13. And each author rethinks the ontological relation of human and object from their specific Native philosophy.

41 At the helm of Kao's global innovation race is Denmark, one of the few countries that has surpassed the United States owing to its evolution from an economy based on agriculture to one based on industry and then on innovation. See Kao, *Innovation Nation.*

42 Kao, *Innovation Nation,* 10–11.

43 See Masco, *Theatre of Operations*; and Parks and Kaplan, *Life in the Age of Drone Warfare.*

44 Dunn, *Militarization of the U.S.-Mexico Border*; and Palafox, "Opening Up Borderland Studies."

45 See Byrd, *Transit of Empire*; LaDuke, *Militarization of Indian Country*; Dunbar-Ortiz, *Indigenous People's History,* 56.

46 Virilio, *War and Cinema,* 4. As a cultural studies scholar, Virilio traces the logics of war alongside cinema, since films mediate how war is perceived by the public.

47 Kaplan, *Aerial Aftermaths.*

48 Nancy Farriss mentions Spanish *vigías,* or coastal lookouts, in her book *Maya Society under Colonial Rule.*

49 A. Smith, "Not-Seeing."

50 Gómez-Barris, The *Extractive Zone,* 6.

51 Wolfe, "Settler Colonialism"; and A. Smith, "Not-Seeing." Scholars have begun to stretch this history further back, to think more centrally about settler colonialism and slavery (which I discuss later), but also to consider how technologies of surveillance were key to extending US imperial optics of control to exterior territories like Cuba and the Philippines. See for example, Alfred W. McCoy, *Policing America's Empire.*

52 A. Smith, "Not-Seeing," 26.

53 Even across an impressive array of scholarship in American studies, science and technology studies, and history, scholars acknowledge the import of the technological "revolution" of modern technologies in asserting the civilizing force of American empire over the primitive other. But the centrality of Native vision-knowledges (Nativision) has not been part of this story.

54 See Benjamin, "Catching Our Breath"; Benjamin, "Introduction."

55 David Lyon refers to the ubiquity of surveillance to describe the pervasiveness of technologies that store data and sort people based on our everyday activities. Critical of the use of the term *surveillance society* as homogenizing the gaze of surveillance, he asks us to consider the particular sites of surveillance that affect some more than others. See Lyon, *Surveillance Studies,* 25.

56 See Pegler-Gordon, *In Sight of America*; Stern, *Eugenic Nation*; and Molina, *Fit to Be Citizens?*

57 Blackfoot philosopher Leroy Little Bear observes, "The human brain is a station on the radio dial; parked in one spot, it is deaf to all the other stations . . . the animals, rocks, trees, simultaneously broadcasting across the whole spectrum of sentience." Quoted in Don Hill, "Listening to Stones: Learning in Leroy Little Bear's Laboratory; Dialogue in the World Outside," *Alberta Views: The Magazine for Engaged Citizens*, September 1, 2008, https://albertaviews.ca/listening-to-stones/.

58 Cajete, *Native Science*. Also see Tambiah, *Magic, Science, Religion*.

59 Participating in a longer colonial resistance movement against occupation, Native Hawaiians raised the issue of environmental destruction of endangered species habitats on the summit, threats to one of the island's aquifers, and the fact that Mauna Kea is a traditional burial ground. For more about this struggle, see Anna Keala Kelly, "Mauna Kea Is Only Latest Thing They Want to Take, 'We Will Not Give It to Them,'" *Indian Country Today*, July 21, 2019, https://indiancountrytoday .com/news/mauna-kea-is-only-latest-thing-they-want-to-take-we-will-not-give-it-to -them?redir=1. Kim Tallbear also discusses the ways Indigenous knowledges are converted into "beliefs" when they protest certain scientific methods such as genome research on their blood or on the bones of their ancestors. See Tallbear, "Beyond the Life/Not-Life Binary," 192.

60 George Johnson, "Seeking Stars, Finding Creationism," *New York Times*, October 20, 2014, https://www.nytimes.com/2014/10/21/science/seeking-stars-finding -creationism.html.

61 Hobart, "At Home on the Mauna," 30.

62 Hobart, "At Home on the Mauna," 31.

63 Brandt, "Fight for Dzil Nchaa Si An," 55.

64 LaDuke, *Recovering the Sacred*, 13.

65 Henry M. Teller, a lawyer and general of the Colorado militia, pushed to pass the 1883 Code of Indian Offenses as a tactic meant to expedite the assimilation of Native peoples and settler control over land. This law not only criminalized various sacred dances and the activity of medicine men and female healers but allowed the federal government to intervene in the "criminal" affairs of sovereign reservations.

66 Basso, *Wisdom Sits in Places*, 154; Laluk, "Indivisibility of Land and Mind"; Watts, "Indigenous Place-Thought."

67 Laluk, "Indivisibility of Land and Mind," 98.

68 L. Green and D. Green, *Knowing the Day*.

69 L. Green and D. Green, *Knowing the Day*, 174.

70 L. Simpson, *As We Have Always Done*. Also see Mohawk and Anishnaabe sociologist Vanessa Watts, who argues that place-thought is "based on the premise that land is alive and thinking and that humans and nonhumans derive agency through the extensions of these thoughts." Watts, "Indigenous Place-Thought," 21.

71 L. Simpson, *As We Have Always Done*, 151.

72 L. Simpson, *As We Have Always Done*, 154.

73 In this essay Whyte is critical of how non-Indigenous scholars transplant TEK into Western knowledge systems: Whyte, "Role of Traditional Ecological Knowledge."

74 Dillon, introduction, 7.

75 V. Deloria, *Spirit and Reason.*

76 Kimmerer, "Weaving Traditional Ecological Knowledge."

77 Cajete, *Native Science*, 40.

78 Peat, *Blackfoot Physics*, 6.

79 See in particular chapter 2, "Andean Phenomenology and New Age Settler Colonialism," in Gómez-Barris, *Extractive Zone.*

80 Willey, "Biopossibility"; Tallbear, "Making Love and Relations."

81 Rifkin, *Erotics of Sovereignty.*

82 Escobar, "Thinking-Feeling with the Earth."

83 Todd, "Refracting the State."

84 Roy, *Molecular Feminisms.*

85 Lopez, "Change on the Amah Mutsun Tribe."

CHAPTER 1. "THE EYES OF THE ARMY"

1 This bronze sculpture was commissioned from a Tucson artist, Dan Bates. The sculpture resembles the style of Fredrick Remington, a well-known artist and sculptor who created paintings and bronze sculptures of the cavalry and Indian wars of the West during the late 1880s. As a gesture of commemoration, Huachuca officers opened the fort to members of the San Carlos and White Mountains Apache tribes during the late 1980s until today. They return to Fort Huachuca to engage in ceremony on their ancestral land.

2 I use *Indian* when referring to the historically specific term used by the cavalry to discuss the Apache and others, *Native American* when describing various individuals and communities in the United States, and *Indigenous* as a more politicized and critical term that refers to a collectively imagined global community.

3 Lt. Col. Patrick Sullivan, commander of the Unmanned Aircraft Systems Training Battalion, says, "Army UAS are the 'Eyes of the Army.'" See Amy Sunseri, "UASTB Largest UAS Training Center, 'Pilots' Unique Mission," U.S. Army, May 20, 2010, https://www.army.mil/article/39475/uastb_largest_uas_training_center_pilots _unique_mission. Ikhana is a NASA drone not on display at Huachuca but relevant here as its name draws on a Native American Choctaw word meaning "intelligent, conscious, or aware." As the first drone to fly in commercial airspace, this drone's unique intelligence allows it to communicate information to other aircraft to avoid air collisions. Daniel Terdiman, "In a First, NASA's Predator Drone Flew Solo in US Commercial Airspace," *Fast Company*, June 12, 2018, https://www .fastcompany.com/40584304/in-a-first-nasa-predator-drone-flew-solo-in-commercial -airspace.

4 Army Intelligence Museum, Fort Huachuca, "Intelligence Impulse," 23.

5 They continue, "[It] can detect people moving up to three miles and can spot vehicles at over six miles, making it useful for detecting enemy movements and provide early warning." Army Intelligence Museum, Fort Huachuca, "Intelligence Impulse," 21.

6 Army Intelligence Museum, Fort Huachuca, "Intelligence Impulse," 21.

7   As argued by Edwin Layton, in the 1900s "technological knowledge was uprooted from its matrix in centuries-old craft traditions and grafted onto science. . . . [T]he oral traditions passed from master to apprentice, the new technologist substituted a college education, a professional organization, and a technical literature patterned on those of science." Layton, "Mirror-Image Twins," 562.

8   Bonds and Inwood, "Beyond White Privilege."

9   Olivia B. Waxman, "'We Became Warriors Again': Why World War I Was a Surprisingly Pivotal Moment for American Indian History," *Time*, November 23, 2018, https://time.com/5459439/american-indians-wwi/.

10   See LaDuke, *Militarization of Indian Country*, 3. Native warriors were celebrated for their efforts in the Civil War, as code talkers (or wind talkers) during World War II, and as brave warriors in the Vietnam, Afghanistan, and Iraq wars. LaDuke wants us to remember the continued role of the military in dumping toxic materials on Native American lands, the military's role in land grabs, the lingering effects of post-traumatic stress disorder, and the countless deaths that may be erased when the military commemorates Native war heroes.

11   LaDuke, *Militarization of Indian Country*, 9.

12   LaDuke also reminds us that when the 1924 Indian Citizenship Act was passed, Native peoples were eligible to be drafted into war, while other rights, such as voting were barred by state laws. LaDuke, *Militarization of Indian Country*, 11.

13   The Chiricahua were a group of different Apache bands who called themselves *Nde or N'dee*, which meant "the people." There were many nations of Apache people at this time including the Mescalero, Jicarrilla, Lipan, and Kiowa-Apache peoples. The term Apache meant "enemy" in Zuni and was used by the Spanish for this reason. See "Apache before 1861," National Park Service, accessed April 2, 2020, https://www.nps.gov/chir/learn/historyculture/pre-apache-wars.htm.

14   Goodwin, *Geronimo*.

15   For more on the Gadsden Purchase, see Schmidt, "Manifest Opportunity."

16   Euro-Americans allowed Native groups as diverse as the Navajo, Apaches, Comanches, Yumas, and others to raid Mexicans in the hopes of pushing the Mexicans farther south. Surveyor accounts repeat their amazement at how many southern Arizona towns, mines, and buildings were left abandoned. See Wallace, *Great Reconnaissance*. Also, the Arizona Organic Act of 1862 split Arizona off from the New Mexico Territory, creating the Arizona Territory and establishing the current boundaries around these states.

17   While most historical accounts describe the Indian scouts as falling under the umbrella of the Apache during the Indian Wars, other groups such as the Sioux were also regarded as excellent scouts.

18   Brian Hochman makes this argument in relation to the rise of ethnography through the urgent project of documenting the dying languages of American Indians during the late nineteenth and early twentieth century. Hochman, *Savage Preservation*.

19   James P. Finley, "Indian Scouts at Huachuca in the 1920s and 1930s," *Huachuca Illustrated*, volume 2, http://www.gwpda.org/comment/huachuca/HI2-25.htm.

From 1993 to 1996, Finley, Fort Huachuca's post historian, wrote and compiled a comprehensive and well-illustrated series *The Buffalo Soldiers at Fort Huachuca*, which formed the first three issues of the historical magazine of Fort Huachuca, called *Huachuca Illustrated: A Magazine of the Fort Huachuca Museum*.

20 Huachuca submitted this historical documentation in 1996 once the National Historic Preservation Act (P.L. 89-665, as amended) passed, turning Fort Huachuca into a site of public interest in addition to a military intelligence school.

21 Dodge, *Hunting Grounds*, 368–69.

22 Wally Brown, "Indian Sign Language," YouTube video, 8:11, posted August 26, 2019, https://www.youtube.com/watch?v=Xh9opFfPKKI.

23 J. Davis, *Hand Talk*, 83.

24 Hochman, *Savage Preservation*, xiii.

25 Myer's wigwag telegraphy transported secret messages through the use of a flag (or a kerosene torch at night). See J. Davis, *Hand Talk*, 35. Myer studied American Indian and Deaf sign language and Morse code, which led to the creation of the telegraph; in fact, his dissertation was titled "A New Sign Language for Deaf-Mutes." For more about Myer, see J. Davis, *Hand Talk*, 36.

26 For an intriguing book on the connections between American Indian sign or gesture language and policies on Deaf education, see the chapter "Savages and Deaf-Mutes: Species and Race" in Baynton, *Forbidden Signs*, 36–55.

27 Mallery, *Sign Language*.

28 Webb, *Great Plains*, 77; Hochman, *Savage Preservation*; and J. Davis, *Hand Talk*.

29 Quoted in W. P. Clark, *The Indian Sign Language*, 12.

30 J. Davis, *Hand Talk*, 82.

31 Quoted in J. Davis, *Hand Talk*, 65.

32 Mooney, "Sign Language." Mallery's book *Sign Language among North American Indians* (1881) relies on his own observations, as well as others', including those of Colonel Dodge, who wrote about the culture, practices, and gesture language of the Plains Indians.

33 Barrett, *Geronimo's Story*, Kindle, loc. 456 of 2001.

34 Cajete, *Native Science*, 40.

35 One could argue here that animals also became more human through the practice of living with human manners of eating and living, altering the ecological rhythm of place.

36 Quoted in Kay Yandell, *Telegraphies*, 41.

37 Silko, *Almanac of the Dead*, 224–25.

38 See the Fort Huachuca website, "Fort Huachuca Schools and Courses," accessed April 2, 2020, http://tdystay.com/fort-huachuca-schools-courses.

39 Army Intelligence Museum, Fort Huachuca, "Intelligence Impulse," 2.

40 Army Intelligence Museum, Fort Huachuca, "Intelligence Impulse," 2.

41 Army Intelligence Museum, Fort Huachuca, "Intelligence Impulse," 5.

42 It is well known that even Hernán Cortés built an army of recruits against the Aztecs by augmenting his small army of Spanish soldiers with thousands of Indian auxiliaries.

43 For a discussion of Roy E. Finkenbine's forthcoming book, tentatively titled "Fugitive Slaves in Indian Country: Crossings and Sanctuaries," see Finkenbine, "The Native Americans Who Assisted the Underground Railroad," History News Network, Columbian College of Arts and Sciences, September 15, 2019, https://historynewsnetwork.org/article/173041.

44 Cornelius C. Smith Jr., *Fort Huachuca*, 253.

45 Bernardo de Galvez, *Instructions for Governing*, quoted in McLachlan and Charles River Editors, *Apache Scouts*, loc. 22-33 out of 867.

46 Wagner, *Organization and Tactics*, 203.

47 Bourke, *An Apache Campaign in the Sierra Madre*, loc. 14 out of 71.

48 Wagner, *Service of Security and Information*, 56 .

49 Wagner found that Apaches in their own country were ideal for infantry patrols who fight from the ground, while the Sioux is "the best type of mounted Indian. He is 'all eyes and ears,' is seldom seen, and may, in fact, be characterized as a perfect scout." In 1876 the army hired 8,000–10,000 Sioux. See Wagner, *Service of Security and Information*, 201. Also, many times false information was shared when trackers misled the cavalry to protect their people.

50 Wagner, *Service of Security and Information*, 20.

51 Wagner, *Service of Security and Information*, 43.

52 Wagner, *Service of Security and Information*, 201.

53 Wagner, *Service of Security and Information*, 3.

54 Wagner, *Organization and Tactics*, 3.

55 Wagner, *Organization and Tactics*, 40.

56 Quoted in Goodwin, *Geronimo* at 25:00.

57 Alan Houser, quoted in Goodwin, *Geronimo* at 25:39.

58 Unremarked was their dark skin color, which allowed them to travel without being seen at night.

59 Barr, *Peace Came*. Also see Goeman, *Mark My Words*.

60 Wagner, *Service of Security and Information*, 204.

61 Barrett, ed. and trans. *Geronimo's Story*, loc. 547 of 2001

62 Melford Yuzos, quoted in Goodwin, *Geronimo* at 6:58.

63 I found this source about Lozen in Lauren Redniss's book *Oak Flat: A Fight for Sacred Land in the American West*, 248. She quotes from James Kaywaykla's oral history in Ball, *In the Days of Victorio*, 11.

64 Quoted in Thrapp, "Indian Scouts."

65 Miles, quoted in Goodwin, *Geronimo* at 32:28.

66 Miles, *Personal Recollection*, 480.

67 Miles, *Personal Recollection*, 480.

68 Miles had read about the use of the heliograph by the British Army in India and had limited experience with this technology during the Civil War. See Smith, *Fort Huachuca*, 122.

69 Gordillo, "Terrain as Insurgent Weapon," 58.

70 Gordillo, "Terrain as Insurgent Weapon," 58.

71 Gordillo, "Terrain as Insurgent Weapon," 56. Also see U.S. Army Intelligence Center of Excellence, Marilyn Willis-Grinder, "Cross-Cultural Competence," *MIPB*

(Military Intelligence Professional Bulletin), January–March 2011, 2-9. https://ufdc.ufl
.edu/AA00062632/00036.

72 Gordillo, "Terrain as Insurgent Weapon," 56.
73 Skelton, "America's Frontier Wars," 23.
74 Skelton, "America's Frontier Wars," 25.
75 Skelton, "America's Frontier Wars," 25.
76 Skelton, "America's Frontier Wars," 26.

CHAPTER 2. OCCUPATION ON SACRED LAND

1 The first black hole was discovered in 1971, spurring lively debates on the force of
black holes to suck everything into their orbit.
2 See Christopher Bolkcom, "Homeland Security: Unmanned Aerial Vehicles and
Border Surveillance," 3.
3 See Oliphant v. Suquamish Indian Tribe, 435 U.S. 191, 194 (1978).
4 Benton, "Spatial Histories of Empire," 29-30.
5 A. Simpson, *Mohawk Interruptus*, 124.
6 See Nevins, *Operation Gatekeeper*; and De León, *Land of Open Graves*. There is also
a discussion of the border security crisis on the O'odham reservation by O'odham
members in *The Impact of the Drug Trade on Border Security and National Parks: Hearing
before the Subcommittee on Criminal Justice, Drug Policy, and Human Resources of the Committee on Government Reform, House of Representatives*, 108th Congress, First Session
(March 10, 2003), https://www.govinfo.gov/content/pkg/CHRG-108hhrg87703
/html/CHRG-108hhrg87703.htm.
7 I use the O'odham measurement of the border. By US accounts, this border region
is seventy-five miles long.
8 This includes recognized and unrecognized nations such as the Navajo, Kickapoo,
Iroquois, and many others.
9 I discuss the colonial fabrication of empty space in the introduction. For a savvy account of quantum understandings of the void as a helpful way to think through the
erasure of nuclear devastation on Native land across time/s, see Barad, "Troubling
Time/s."
10 Marcelo Di Cintio, "Ofelia Rivas, the Tohono O'odham, and the Wall," January 19,
2019, https://marcellodicintio.com/2019/01/19/ofelia-rivas-the-tohono-oodham-and
-the-wall/.
11 D. Saxton, L. Saxton, and S. Enos, *Tohono O'odham/Pima Dictionary*, 22.
12 Thank you, Iriany, for introducing me to such amazing friends from the O'odham
reservation.
13 A. Simpson, *Mohawk Interruptus*, 116. Simpson is referring to scholars such as Gloria
Anzaldúa and many others influenced by her work.
14 B. Anderson, "More Equal Than Others."
15 A. Smith, *Conquest*, 182.
16 Fernanda Santos, "Border Wall Would Cleave Tribe and Its Connection to Ancestral Land," *New York Times*, February 20, 2017, https://www.nytimes.com/2017/ 02/20
/us/border-wall-tribe.html.

17 De Genova, "Denizenship," 231.

18 De Genova, "Denizenship," 231.

19 Mezzadra and Neilson, *Border as Method*.

20 Numerous media, scholarly, photographic, and artistic accounts of the desert border region of Arizona depict this land as vacant, or unpopulated, except for the deceased remnants of migrants coming from Mexico and Latin America. This is true in particular of Jason De León's *The Land of Open Graves*, but even activist border artworks and organizations omit Native Americans from the land.

21 Jodi Byrd uses the term *arrivant* in *Transit of Empire* to consider the complicated position of groups such as the formerly enslaved or the indentured who did not voluntarily come to North America. I would say that choice for most migrants is a complex concept when so many are fleeing persecution, violence, and rampant poverty. Also see Tuck and Yang, "Decolonization Is Not a Metaphor." I am also thinking of Maylei Blackwell, Floridalma Boj Lopez, and Luis Urrieta Jr.'s important special issue of *Latino Studies*, titled "Critical Latinx Indigeneities."

22 See Anzaldúa, *Light in the Dark*; and Schaeffer, "Spirit Matters."

23 This website is no longer active, but many articles are now available at *Censored News: Indigenous Peoples and Human Rights*, July 30, 2013, https://bsnorrell.blogspot .com/2013/07/oodham-voice-against-wall-us-border.html

24 See Simmons, "Settler Atmospherics."

25 In "Decolonization Is Not a Metaphor," Eve Tuck and K. Wayne Yang remind scholars that "land" includes water, land, air, and subterranean earth. Aerial sovereignty is often absent from discussions, except among scholars such as Simmons who consider the toxic fumes of environmental waste.

26 Scientists amassed volumes of data on the flora and fauna of the border in geological and paleontological reports on the region and sketches of the natural wonders of the area. This is detailed in Wallace, *Great Reconnaissance*.

27 Mishuana Goeman references Sacagawea, one of the most famous female guides, whose knowledge was formative to the 1804 Lewis and Clark expedition yet was absented from the maps. Goeman, *Mark My Words*, 24.

28 Wallace, *Great Reconnaissance*, 94.

29 As argued by Geraldo Cadava, "Article 11 of the Treaty of Guadalupe Hidalgo singled out Native Americans and obligated the United States to prevent their raids into Mexico, even though U.S. filibusters and 'white Cowboys' raided as well." See Cadava, "Borderlands of Modernity," 365.

30 The O'odham lost a great deal of land during the second half of the nineteenth century (both in the United States and Mexico as both nations privatized land during this time) after the United States passed the 1862 Homestead Act and 1877 Desert Land Act to spur economic development. See Schulze, *Are We Not Foreigners Here?*, loc. 1191 of 6843, Kindle.

31 P. Deloria, *Indians in Unexpected Places*, 27.

32 Caitlin Blanchfield and Nina Valerie Kolowratnik, "At the Border: Significant Impact," *E-flux Architecture*, April 22, 2020, https://www.e-flux.com/architecture/at-the -border/325749/significant-impact/.

33   The uneven practice of leasing out personal land, or tribally constituted land, to outside entities created rifts in O'odham communities. See Meeks, *Border Citizens*, 33, 55–56.

34   Schulze, *Are We Not Foreigners Here?*, loc. 1250 of 6843, Kindle. In the conclusion I discuss some of the ways O'odham land has been diminished over time.

35   Wolfe, "After the Frontier," 34.

36   Many other acts were passed to encourage non-Native settlement in the West, such as the Desert Land Act of 1877. Sacred regions of O'odham land, such as the area around the Salt River, were taken.

37   Schulze, *Are We Not Foreigners Here?*

38   Schulze, *Are We Not Foreigners Here?*, ch. 5.

39   Without proper documentation, Mexico did not recognize their right to the land.

40   USA Today Network, "The Wall: Unknown Stories. Unintended Consequences," 2017, accessed April 20, 2020, https://www.usatoday.com/border-wall/.

41   This act was supported by congresspeople on both sides of the political line, including Senators Hillary Clinton and Barack Obama. Only the wealthy who could afford lawyers were fairly compensated.

42   USA Today Network, Dennis Wagner, "A 2,000-mile Journey in the Shadow of the Border Wall," 2017, accessed April 20, 2021, https://www.usatoday.com/border-wall/story/flight-over-entire-us-mexico-border-fence/605855001/.

43   See the special series called "The Taking" that includes articles by ProPublica reporter T. Christian Miller and *Texas Tribune* reporters Kiah Collier and Julián Aguilar, especially "The Taking: How the Federal Government Abused Its Power to Seize Property for a Border Fence," *Texas Tribune*, December 14, 2017, https://www.texastribune.org/2017/12/14/border-land-grab-government-abused-power-seize-property-fence/. James Baldwin named evictions in San Francisco "Negro removal" in a 1963 television interview posted on YouTube on June 3, 2015, https://www.youtube.com/watch?v=T8Abhj17kYU.

44   Caitlin Blanchfield and Nina Valerie Kolowratnik, "Assessing Surveillance: Infrastructures of Security in the Tohono O'odham Nation," Archinect, February 13, 2018, https://archinect.com/features/article/150049769/assessing-surveillance-infrastructures-of-security-in-the-tohono-o-odham-nation.

45   "Dakota Access Pipeline Opponents Occupy Land, Citing 1851 Treaty," *Reuters*, October 24, 2016, https://www.reuters.com/article/us-usa-pipeline-dakotaaccess/dakota-access-pipeline-opponents-occupy-land-citing-1851-treaty-idUSKCN12O2FN.

46   Schulze, *Are We Not Foreigners Here?*, loc. 3493 of 6843, Kindle.

47   Byron Pitts and Dan Lieberman, "In Efforts to Secure US-Mexico Border, Arizona Native Americans Feel Caught in the Middle," ABC News, June 27, 2013, http://abcnews.go.com/US/efforts-secure-us-mexico-border-ariz-nativeamericans/story?id=19496394.

48   Pitts and Lieberman, "In Efforts to Secure US-Mexico Border, Arizona Native Americans Feel Caught in the Middle."

49   Robert J. Lopez, Richard Marosi, and Rich Connell, "A 'Black Hole' on a Porous Border," *Los Angeles Times*, May 21, 2006, http://articles.latimes.com/2006/may/21

/local/me-border21. This article specifically discusses a small Mexican village called Jacume. More recently, Trump used this same language, as can be seen in the article by Nancy Bilyeau, "Trump Border Crackdown Opens 'Black Hole' in Protection for Immigrants," *Crime Report*, October 10, 2018, https://thecrimereport.org/2018/10/10 /trump-border-crackdown-opens-black-hole-in-protection-for-immigrants/.

50 Border enforcement as a field of reality television is pervasive in the United States but also around the globe. James Walsh discusses border reality shows in Australia but notes it as a global phenomenon. Walsh, "Border Theatre." Also, numerous reality docudrama television shows in the United States take viewers to see actual encounters with migrants, smugglers, and gangs moving across the border. For the more recent ones, see the National Geographic's *Border Wars* (2010) and the Discovery Channel's *Border Live* (2018). There are also scores of films and documentaries on the border, including a recent one called *Desierto* (2015) that follows an Anglo vigilante who spends the entire film hunting down migrants crossing the desert.

51 Diana M. Náñez, USA Today Network, 2017, accessed April 20, 2021, https://www .usatoday.com/border-wall/story/tohono-oodham-nation-arizona-tribe/582487001/.

52 Ofelia Rivas, "O'odham VOICE against the WALL: U.S. Border Patrol Violates O'odham Rights," now available at *Censored News: Indigenous Peoples and Human Rights*, July 30, 2013, https://bsnorrell.blogspot.com/2013/07/oodham-voice-against -wall-us-border.html.

53 The purpose of the declaration was to establish a set of individual and collective rights including cultural rights and identity rights to education, health, language, and so on.

54 Rivas, "O'odham VOICE against the WALL."

55 Tom Boswell, "Caught in the Crossfire: Border Crisis Threatens Traditional Way of Life for Sovereign Tohono O'odham Nation," *Meditations on the Collapse*, January 21, 2011, http://meditationsoncollapse.blogspot.com/2011/01/caught-in-crossfire-border -crisis.html.

56 Quoted in Stephanie Innes, "Tohono O'odham Leaders: Trump's Wall Won't Rise on Tribal Borderland," *Arizona Daily Star*, February 20, 2017, http://www.tonation -nsn.gov/wp-content/uploads/2017/02/Tohono-Oodham-Leaders-Trumps-Wall -Wont-Rise-on-Tribal-Borderland.pdf.

57 Schulze, *Are We Not Foreigners Here?* This trend has only continued. A 2005 report states that three of the largest communities on the Tohono O'odham reservation are 41 percent to 50 percent below the national poverty level. See Arizona Department of Health Services, *Arizona Health Status*, 433–34.

58 See Audra Mitchell, "Decolonizing against Extinction Part II: Extinction Is Not a Metaphor—It Is Literally Genocide," *Worldly*, December 27, 2017, https://worldlyir .wordpress.com/2017/09/27/decolonizing-against-extinction-part-ii-extinction-is -not-a-metaphor-it-is-literally-genocide/. Other scholars such as Lauren Berlant and Jasbir Puar have usefully expanded the temporality of violence through concepts such as "slow death" and "disabling." See Berlant, "Slow Death"; Puar, *Right to Maim*.

59 Mitchell quotes from Tasha Hubbard's article "Buffalo Genocide in Nineteenth-Century North America: 'Kill, Skin, and Sell.'"

60 See page 113 of O'odham member and scholar Robert Cruz's "AM TÑE'OK ET A:T O CE:EKT DO'IBIODA:LIK"/"In Our Language Is Where We Will Find Our Liberation."

61 Brenda Norrell, "U.S. Israeli Pact Targets Traditional Tohono O'odham with 15 New Spy Towers," *Censored News: Indigenous Peoples and Human Rights*, September 7, 2015, https://bsnorrell.blogspot.com/2015/09/us-israeli-pact-targets-traditional.html.

62 US Department of Homeland Security, *Environmental Assessment*.

63 Rivas, "O'odham VOICE against the WALL."

64 See Caitlin Blanchfield, Nina Valerie Kolowratnik, and Ophelia Rivas's report and map, "Effects of Integrated Fixed Towers," September 22, 2017, accessed July 14, 2020, https://web.archive.org/web/20200611054743/http://www.solidarity-project .org/.

65 Blanchfield, Kolowratnik, and Rivas, "Effects of Integrated Fixed Towers."

66 Blanchfield and Kolowratnik, "At the Border."

67 Also see Blanchfield and Kolowratnik, "'Persistent Surveillance.'"

68 Peter Hardin, "Eyes in the Skies," *Richmond Times-Dispatch*, October 30, 2003, F1.

69 Escobar quotes here from Varela's book *Ethical Know-How: Action, Wisdom and Cognition*. See Escobar, *Designs for the Pluriverse*, 165.

70 Gregory, "Dirty Dancing," 45.

71 Munro, "Mapping the Vertical Battlespace," 238.

72 See Cyndy Cole, "Hualapai, Pilot at Stalemate," *Arizona Daily Sun*, February 17, 2009, http://azdailysun.com/news/hualapai-pilot-at-stalemate/article_f1 7b9b9c -9ceb-53e5-b6eb-aa5524193aea.html. For a discussion of the legal parameters of aerial sovereignty on Native American reservations, see Haney, "Protecting Tribal Skies."

73 Their drones were shot down on their own land. See the video "Meet the Drone Operators of #NoDAPL," December 2, 2016, Facebook video, 6:03, https://www .facebook.com/watch/?v=850063135135195.

74 See Alleen Brown, Will Parrish, and Alice Speri, "Police Used Private Security Air-craft for Surveillance in Standing Rock No-Fly Zone," *Intercept*, September 29, 2017, https://theintercept.com/2017/09/29/standing-rock-dakota-access-pipeline-dapl-no -fly-zone-drones-tigerswan/. For more about the no-fly zone in Ferguson, see The Associated Press, "Police Targeted Media With No-Fly Zone over Ferguson, Tapes Show," *The New York Times*, November 2, 2014, https://www.nytimes.com/2014/11/03 /us/police-targeted-media-with-no-fly-zone-over-ferguson-tapes-show.html.

75 Ofelia Rivas, "Borderlands | Bioneers," presentation as part of the Indigeneity Forum at the 2019 National Bioneers Conference, YouTube video, 13:21, posted March 4, 2020, https://www.youtube.com/watch?v=0E6qlgkNsFs.

76 Boswell, "Caught in the Crossfire."

77 See "History and Culture," Official Website of the Tohono O'odham Nation, 2016, last accessed August 1, 2021, http://www.tonation-nsn.gov/history-culture/. Rivas herself has experienced officials who threatened to deport her if she did not carry the appropriate ID cards while crossing onto the Mexican side of the reservation.

78 Private conversation (in a mix of English and Spanish) with an O'odham resident off the reservation, November 11, 2018.

79 This is not always the case, although it is true that many non-Indigenous aid workers have not suffered the same criminalization as O'odham until more recently. For
a recent case, see Amy Goodman's story on *Democracy Now!* about an activist facing
twenty years in prison for providing food and water to migrants. "Scott Warren
of No More Deaths Faces Retrial for Providing Humanitarian Aid to Migrants in
Arizona," *Democracy Now!*, July 3, 2019, video, https://www.democracynow.org/2019
/7/3/no_more_deaths_scott_warren_retrial.

80 Ricardo Dominguez, an associate professor at the University of California, San
Diego, and a team of researchers faced criminal charges while under an investigation by the Federal Bureau of Investigations (FBI) and the University of California Office of the President (UCOP) for their humanitarian project called "The
Transborder Immigrant Tool Project." All charges were ultimately dropped for their
online art project that added poetry to a GPS device installed on a cell phone to
help migrants locate water as they cross the desert between Mexico and California.
For more about the project, see Leila Nadir, "Poetry, Immigration and the FBI: The
Transborder Immigrant Tool," https://hyperallergic.com/54678/poetry-immigration
-and-the-fbi-the-transborder-immigrant-tool/.

81 *The Impact of the Drug Trade on Border Security: Hearing before the Subcommittee on
Criminal Justice, Drug Policy, and Human Resources of the Committee on Government
Reform, House of Representatives*, 108th Congress, Second Session (April 15, 2003),
https://www.govinfo.gov/content/pkg/CHRG-108hhrg90205/html/CHRG
-108hhrg90205.htm.

82 Randal C. Archibold, "In Arizona Desert, Indian Trackers vs. Smugglers," *New York
Times*, March 7, 2007, https://www.nytimes.com/2007/03/07/washington/07wolves
.html.

83 Brian Bennett, "Indian 'Shadow Wolves' Stalk Smugglers on Arizona Reservation,"
Los Angeles Times, November 21, 2011, https://www.latimes.com/world/la-xpm-2011
-nov-21-la-na-adv-shadow-wolves-20111122-story.html.

84 Bennett, "Indian 'Shadow Wolves.'"

85 Bennett, "Indian 'Shadow Wolves.'"

86 Mark Wheeler, "Shadow Wolves," *Smithsonian Magazine*, January 2003, https://www
.smithsonianmag.com/travel/shadow-wolves-74485304/.

87 Dianna M. Náñez, "A Border Tribe, and the Wall That Will Divide It," from "The
Wall," *USA Today* 2017, accessed March 20, 2020, https://www.usatoday.com/border
-wall/story/tonhono-oodham-nation-arizona-tribe/582487001.

88 Wheeler, "Shadow Wolves."

89 Wheeler, "Shadow Wolves." Also see Emily Cosenza, "Native American Trackers to Hunt Bin Laden," *Australian*, March 12, 2007, https://www.theaustralian
.com.au/national-affairs/defence/native-american-trackers-to-hunt-bin-laden
/news-story/ea5ea1f7c3aa94353ad6546494623c97; and Eugen Tomiuc, "Moldova:
Native American 'Shadow Wolves' Helping Train Moldovan Guards to Protect
Borders," *Radio Free Europe/Radio Liberty*, October 5, 2004, https://www.rferl.org/a
/1055166.html.

90 Archibold, "In Arizona Desert, Indian Trackers."

91 John Ahni Schertow, "UN Asked to Stop Guns at Akwesasne Border," *Interconti-nental Cry*, May 28, 2009, https://intercontinentalcry.org/un-asked-to-stop-guns-at-akwesasne-border/.

92 According to Tom Boswell of the *National Catholic Reporter*, "Per capita income hovers around $8,100 a year, lowest among all U.S. reservations. . . . The Arizona Department of Commerce reported that the unemployment rate for the Tohono O'odham Nation was just over 35 percent this year, but the 2005 American Indian Population and Labor Force Report measured it at 75 percent." Boswell, "Caught in the Crossfire." For a comparison of the effects of drug trafficking on the Tohono O'odham and Mohawk nations, see Revels and Cummings, "Impact of Drug Trafficking."

93 Boswell, "Caught in the Crossfire."

94 Ofelia Rivas, "Immigration, Imperialism, and Cultural Genocide: An Interview with O'odham Activist Ofelia Rivas Concerning the Effects of a Proposed Wall on the U.S.—Mexico Border," interview by Jeff Hendricks, Tiamat Publications, July 2003, accessed December 11, 2018, https://online.fliphtml5.com/grfc/mhhf/#p=1.

95 Rivas, "Immigration, Imperialism, and Cultural Genocide."

96 Rivas, "Immigration, Imperialism, and Cultural Genocide."

97 Ramirez, *Native Hubs*, 69.

98 Quoted in Náñez, "Border Tribe."

CHAPTER 3. AUTOMATED BORDER CONTROL

1 In 2012 the DHS tested iris recognition remotely on immigrants in Texas, and currently border agents run facial recognition scans on travelers with visas after President Donald Trump declared a biometric war in January 2017 against those who might overstay their visas. Adam Clark Estes, "Coming to an Airport Near You: Iris Scanners," *The Atlantic*, June 7, 2011, http://www.theatlantic.com/tech-nology/archive/2011/06/coming-airport-near-you-iris-scanners-351530/; Jacob Goodwin, "Iris Recognition Is Being Tested on Illegal Immigrants in McAllen, TX," *Government Security News*, August 31, 2012, accessed April 22, 2018, http://www.gsnmagazine.com/node/27136 (no longer available); and Ron Nixon, "Border Agents Test Facial Scans to Track Those Overstaying Visas," *New York Times*, August 1, 2017, https://www.nytimes.com/2017/08/01/us/politics/federal-border-agents-biometric-scanning-system-undocumented-immigrants.html.

2 Inspired by the Silicon Valley in northern California, Robert Breault created the Arizona Optics Industry Association, which has attracted hundreds of companies to Tucson to develop "lasers, telescopes, endoscopy machines and camera lenses, all of which use light, or optics, to enhance images." See Silvana Ordoñez, "Meet Robert Breault, the Businessman behind Tucson's 'Optics Valley,'" *Arizona Daily Star*, January 22, 2012, https://tucson.com/business/local/meet-robert-breault-the-businessman-behind-tucson-s-optics-valley/article_3f6a1c99-ced0-5e76-b348-eod098dc32b7.html.

3 Foucault, *History of Sexuality*, 133.

4   Gates, *Our Biometric Future*; Magnet, *When Biometrics Fail*.

5   US Department of Justice, "Identity Theft," 1.

6   In his discussion of the "securitization of identity," Nikolas Rose contends that the exercise of freedom and active citizenship require legitimate proof of one's identity. See Rose, *Powers of Freedom*, 34.

7   Janet Napolitano, *Statement for the Record: The Honorable Janet Napolitano, Secretary, United States Department of Homeland Security, before the United States Senate Homeland Security and Governmental Affairs Committee* (April 17, 2013), 2, https://www.hsgac .senate.gov/imo/media/doc/Testimony-Napolitano-2013-04-17-REVISED.pdf.

8   Will Parrish, "The U.S. Border Patrol and an Israeli Military Contractor Are Putting a Native American Reservation under 'Persistent Surveillance,'" *Intercept*, August 25, 2019, https://theintercept.com/2019/08/25/border-patrol-israel-elbit -surveillance/.

9   Hank Stephenson, "Are We Telling the Truth? Ask AVATAR!," *Nogales International*, February 10, 2011, https://www.nogalesinternational.com/news/are-we-telling-the -truth-ask-avatar/article_607c949f-db47-5b38-87fd-38ebca3c40c5.html.

10 This was also argued by Simone Browne in "Digital Epidermalization."

11 Golash-Boza, "Mass Incarceration and Mass Deportation."

12 Chacón and Davis, *No One Is Illegal*, 222–23.

13 Massey, "How Arizona Became Ground Zero."

14 See sections 212, 215, and 218 of the Illegal Immigration Reform and Immigrant Responsibility Act of 1996, Division C, Pub. L. No. 104-208, 8 U.S.C. (1996).

15 Byron Pitts and Dan Lieberman, "In Efforts to Secure US-Mexico Border, Arizona Native Americans Feel Caught in the Middle," ABC News, June 27, 2013, http:// abcnews.go.com/US/efforts-secure-us-mexico-border-ariz-nativeamericans/story?id =19496394.

16 On President Trump's border-wall expenditures, see David J. Bier, "GOP Bill Spends More on Border Patrol in 5 Years Than It Has Spent in 5 Decades," *Cato at Liberty* (blog), January 24, 2018, https://www.cato.org/blog/gop-bill-spends-more-border -patrol-5-years-it-has-spent-5-decades.

17 Parrish, "U.S. Border Patrol."

18 Golash-Boza, "Mass Incarceration and Mass Deportation," 486.

19 "National Center for Border Security and Immigration," Eller College of Management, University of Arizona, accessed March 21, 2021, https://eller.arizona.edu /departments-research/centers-labs/border-security-immigration.

20 Adam Higginbotham, "Deception Is Futile When Big Brother's Lie Detector Turns Its Eyes on You," *Wired*, January 17, 2013, https://www.wired.com/2013/01/ff-lie-detector/.

21 See Lyon, "Under My Skin," 299.

22 Dijstelbloem, Meijer, and Besters, "Migration Machine," 11. Surveillance scholars argue that biometrics bind identity to the body. See Amoore, "Biometric Borders"; Lyon, "Under My Skin," 306; Cole, *Suspect Identities*; Gates, *Our Biometric Future*; Magnet, *When Biometrics Fail*; Pugliese, *Biometrics*; Ajana, "Recombinant Identities"; Ajana, *Governing through Biometrics*; Adey, "Facing Airport Security."

23 See Cuvier, *Lectures on Comparative Anatomy*; Nott and Glidden, *Types of Mankind*.

24 For a discussion of the history of the term algorithmic governance, see Katzenbach, "Evolving Digital Society."

25 Bianchini, "La mentalità della razza calabrese," 17–18, quoted in Pugliese, *Biometrics*, 84.

26 Regnault, "Le Lange par gestes," 315, quoted in Rony, *Third Eye*, 56–57.

27 Cole, *Suspect Identities*, 20.

28 Lee, "Enforcing the Borders," 62–63.

29 Molina, *Fit to Be Citizens?*

30 See Stern, *Eugenic Nation*.

31 Fassin, "Compassion and Repression," 371–72. Fassin discusses the term *biolegitimacy* through the shift in French asylum cases during the 1990s when the state turned to the body, and in particular illness, as a more authentic claim to asylum than political claims to harm. See also Gündogdu, *Rightlessness*. Btihaj Ajana discusses similar practices for asylum seekers in "Recombinant Identities: Biometrics and Narrative Bioethics."

32 Browne, *Dark Matters*, 7.

33 Quoted in Pugliese, *Biometrics*, 87.

34 Office of the Under Secretary of Defense for Acquisition, Technology, and Logistics, *Report*.

35 Santa Ana, *Brown Tide Rising*; Chavez, *Latino Threat*; and Fregoso, *MeXicana Encounters*.

36 Here I extend Foucault's term *massifying*, with which he thinks about the problem of governing at the level not of the body but of the species or population. I want to think about the biomass as a racializing force governed through fear and thus knowable as a mass threat without the possibility for individuality. For more on the idea of massifying, see Foucault, *"Society Must Be Defended."*

37 Merriam-Webster, s.v. "hot spot (n.)," accessed November 12, 2020, http://www.merriam-webster.com/dictionary/app.

38 For a similar perspective via drone vision, see Parks, "Drones, Infrared Imagery, and Body Heat."

39 BORDERS received a $15 million grant for 2008–15.

40 Moreover, as many scholars have noted, diminished state funding for research has facilitated the rise of neoliberalism in universities—and also of militarization, as the Department of Defense and DHS spread their educational empire across universities, corporations, and labor industries. This fact was not lost on those of us in the University of California as Janet Napolitano moved from her position as first the governor of Arizona and then the head of the DHS to become the president of the university system.

41 Ekman, *Telling Lies*.

42 Ekman and Friesen, "Nonverbal Leakage."

43 Ekman and Friesen, "Nonverbal Leakage." Also see Ekman, *Telling Lies*, 19.

44 Ekman and Friesen, "Nonverbal Leakage."

45 Adey, "Facing Airport Security"; Burgoon, Guerrero, and Mansov, "Nonverbal Signals"; Ekman, *Telling Lies*.

46 Ekman wrote the introduction and conclusion for the 1998 edition of Charles Darwin's book *The Expression of the Emotions in Man and Animals*. Following *The Descent of Man*, Darwin moved away from popular religious beliefs that posited human

expressions were given to us by the creator to communicate intimate feelings. Instead, Darwin studied humans and animals to argue that the similarity of their emotional responses demonstrated that emotions were universal, were passed down through heredity, and thus were clues to the evolutionary puzzle that proved the link between humans and other animals.

47 Burgoon et al., "Patterns of Nonverbal Behavior."

48 Langhals, Burgoon, and Nunamaker, "Eye and Head Based Psychophysiological Cues," 34.

49 Marc Andrejevic argues convincingly that the glut of information is meant to overwhelm us. Furthermore, because such large amounts of data are more accurately deciphered by computational algorithms, humans lose the power to interpret the world. His analysis raises ethical and political dilemmas of posthuman analysis that may collude with the political right to find human testimony, interpretation, and narration to be fallible, more suited to producing fiction than hard scientific results. See Andrejevic, *Infoglut*, 2–3.

50 Higginbotham, "Deception Is Futile."

51 Burgoon, Guerrero, and Mansov, "Nonverbal Signals"; Ekman, "Facial Expression and Emotion."

52 G. Carte and E. Carte, *Police Reform*, 49–50.

53 Quoted in Matt Stroud, "Will Lie Detectors Ever Get Their Day in Court Again?," *Bloomberg*, February 2, 2015, https://www.bloomberg.com/news/articles/2015-02-02/will-lie-detectors-ever-get-their-day-in-court-again-.

54 Elena, interview with author, University of Arizona, March 25, 2015.

55 Quoted in Johnny Cruz, "UA to Co-lead DHS Center for Border Security and Immigration," University of Arizona, February 26, 2008, https://uanews.arizona.edu/story/ua-to-co-lead-dhs-center-for-border-security-and-immigration.

56 Elena, interview.

57 Elena, interview.

58 Jennifer Terry makes a similar argument in *Attachments to War: Biomedical Logics and Violence in Twenty-First-Century America*.

59 See Conor Friedersdorf, "The NYPD Is Using Mobile X-Ray Vans to Spy on Unknown Targets," *Atlantic*, October 19, 2015, https://www.theatlantic.com/politics/archive/2015/10/the-nypd-is-using-mobile-x-rays-to-spy-on-unknown-targets/411181/.

60 Todd Miller, "How Border Patrol Occupied the Tohono O'odham Nation," *In These Times*, June 12, 2019, https://inthesetimes.com/article/us-mexico-border-surveillance-tohono-oodham-nation-border-patrol. Also see Parrish, "U.S. Border Patrol."

61 Klein, *Shock Doctrine*.

62 Naomi Klein, "Laboratory for a Fortressed World," *Nation*, June 14, 2007, https://www.thenation.com/article/laboratory-fortressed-world/.

63 Klein, *Shock Doctrine*, 550.

64 Todd Miller and Gabriel Matthew Schivone, "Why Is an Israeli Defense Contractor Building a 'Virtual Wall' in the Arizona Desert?," *Nation*, January 26, 2015, https://www.thenation.com/article/why-israeli-defense-contractor-building-virtual-wall-arizona-desert/.

65 Todd Miller and Gabriel M. Schivone, "Gaza in Arizona: The Secret Militarization of the U.S.-Mexico Border," *Salon*, February 1, 2015, https://www.salon.com/2015/02/01/gaza_in_arizona_the_secret_militarization_of_the_u_s_mexico_border_partner/. Also see American Civil Liberties Union, "The Constitution in the 100-Mile Border Zone," accessed March 20, 2015, https://www.aclu.org/other/constitution-100-mile-border-zone.

66 See Nelly Jo David's Twitter hashtag (@USCPR), "Nellie Jo David of Tohono O'odham Hemajkam Rights Network is on the ground in Palestine with powerful reflections on a #WorldWithoutWalls." David is one of the O'odham scholars and activists from the Tohono O'odham Hemajkam Rights Network who accompanied this delegation in solidarity with Palestinians. US Campaign for Palestinian Rights, October 13, 2017, https://twitter.com/uscpr_/status/918901625716080642?lang=en.

67 "'Eyesight' Software Will Scan for Suspicious Behavior," Israeli Homeland Security website, September 9, 2014, http://i-hls.blogspot.com/2014/09/eyesight-software-will-scan-for.html.

68 See Jonathan Karp and Laura Meckler, "Which Travelers Have 'Hostile Intent'? Biometric Device May Have the Answer," *Wall Street Journal*, August 14, 2006, https://www.wsj.com/articles/SB115551793796934752.

69 See Jasbir Puar's book *The Right to Maim: Debility, Capacity, Disability.*

70 Greg Grandin, "How the U.S. Weaponized the Border Wall," *Intercept*, February 10, 2019, https://theintercept.com/2019/02/10/us-mexico-border-fence-history/.

71 De León, *Land of Open Graves.*

72 Moshe Glantz, "Fighting Terrorists with Technology," *Ynetnews*, May 22, 2016, https://www.ynetnews.com/articles/0,7340,L-4806369,00.html. This one includes a video that says, "16 Terrorists with no weapons or explosives changed the world."

73 Glantz, "Fighting Terrorists with Technology."

74 Adey, "Facing Airport Security."

75 Adey, "Facing Airport Security"; Ekman and Friesen, *Facial Action Coding System*; Ekman, Friesen, and Scherer, "Body Movement and Voice Pitch."

76 Jana Winter, "Exclusive: TSA 'Behavior Detection' Program Targeting Undocumented Immigrants, Not Terrorists," *Intercept*, April 6, 2015, http//firstlooklorg/theintercept/2015/04/06/exclusive-tsa-behavior-detection-program-targeting-immigrants-terrorists/.

77 Burns, "Privacy Impact Assessment," 1–2.

78 Middleton, "Privacy Impact Assessment Update," 3.

79 Burns, "Privacy Impact Assessment," 2.

80 For more on the failure of the Boeing surveillance project at the US-Mexico border, see Michael Krigsman, "Boeing Virtual Fence: $30 Billion Failure," *ZDNet*, August 23, 2007, https://www.zdnet.com/article/boeing-virtual-fence-30-billion-failure/.

81 See Steven Levy, "Inside Palmer Luckey's Bid to Build a Border Wall," *Wired*, June 11, 2018, https://www.wired.com/story/palmer-luckey-anduril-border-wall/.

82 Flynn, Juergens, and Cantrell, "Employing ISR," 59.

83 Chow, *Age of the World Target*, 31.

84 Andrea Miller considers how "incitement to violence discourse functions as a technology of statecraft whereby affective realms typically characterized by interiority and unknowability are rendered actionable and criminal for racialized Muslim and Arab bodies in the war on terror." See A. Miller, "(Im)Material Terror," 113.

85 See IBIA (International Biometrics and Identity Association), "Biometrics and Identity in the Digital World," March 17, 2013, https://www.ibia.org/download/datasets/969/Biometrics%20and%20Identity%20in%20the%20Digital%20World.pdf.

86 *Written Statement of James Benjamin Hutchinson, Testimony on Behalf of the International Biometrics + Identity Association (IBIA), before the Committee on Oversight and Government Reform, United States House of Representatives, Law Enforcement's Use of Facial Recognition Technology* (March 22, 2017), 5, https://www.ibia.org/download/datasets/3781/Hutchinson-IBIA-Statement-FRT-Study-3-22.pdf.

87 Benjamin, "Catching Our Breath."

88 See O'Neil, *Weapons of Math Destruction*, 136.

89 Puar, "Precarity."

90 Human Rights Watch, *Forced Apart*. Citing the 1993 World Trade Center bombing and the 1995 Oklahoma City bombing, the Antiterrorism and Effective Death Penalty Act passed through Congress with ease as an antiterrorism measure. It paved the way for IIRIRA, which also prioritized migrant detection by funding border-surveillance technologies at the US-Mexico border.

91 Candice Bernd, "Biden Is Rejecting Trump's Border Wall—But Favors His Own Technological Wall," *Truthout*, February 2, 2021, https://truthout.org/articles/biden-is-rejecting-trumps-border-wall-but-proposing-his-own-virtual-wall/?fbclid=IwAR0bTA_1jXsd7w7bkz03MHTaZCELLrmWbFXxeu1SrPtoJqxXxfLviOirFVg.

92 The Trump administration pressured the Centers for Disease Control and Prevention (CDC) to pass an order in March 2020 that would bypass US and International laws set up to protect refugees and asylum seekers. See Human Rights Watch, "Q&A: US Title 42 Policy to Expel Migrants at the Border," April 8, 2021, https://www.hrw.org/news/2021/04/08/qa-us-title-42-policy-expel-migrants-border#. For an article about the spread of COVID-19 in detention centers and jails see, Joe Walsh, "Report: DHS Admits ICE's Policies Fueled COVID-19 Outbreaks in Jails," October 7, 2020, https://www.forbes.com/sites/joewalsh/2020/10/06/report-dhs-admits-ices-policies-fueled-covid-19-outbreaks-in-jails/?sh=7a909e805581.

93 Nick Statt, "Peter Thiel's Controversial Palantir Is Helping Build a Coronavirus Tracking Tool for the Trump Administration," April 21, 2020, https://www.theverge.com/2020/4/21/21230453/palantir-coronavirus-trump-contract-peter-thiel-tracking-hhs-protect-now.

94 It is perhaps no surprise that a highly marketable and profitable genre of personal biographical stories by migrants has flourished since the 1990s. The majority of these narratives demonstrate migrants' deservingness as entrepreneurial and hardworking. While these stories are important corrections to conservative accusations and accelerating surveillance and laws against the assumed threat of migrant crossings, Sujatha Fernandes argues that "curated personal stories shift the focus away from structurally defined axes of oppression and help to defuse the confron-

tational politics of social movements." Fernandes, *Curated Stories*, 3. I would also say that these accounts have proliferated in response to surveillance that evacuates testimony from the adjudication of migrants or anyone suspected of terrorism.

95 Thanks to Sylvanna Falcón for introducing me to this artist. For artwork by Victor Delfín, see "Inside the Artist's Studio: A Visit with Victor Delfín," Goshen College, March 25, 2014, https://www.goshen.edu/peru/2014/03/25/inside-the-artists-studio-a -visit-with-victor-delfin/.

CHAPTER 4. FROM THE EYES OF THE BEES

Epigraph: Zach Campbell, "Swarms of Drones, Piloted by Artificial Intelligence, May Soon Patrol Europe's Borders," *Intercept*, May 11, 2019, https://theintercept.com /2019/05/11/drones-artificial-intelligence-europe-roborder.

1 The US government has already tested swarm drones in California and deployed them in 2014 to protect manned ships against attack. Todd South, "Drone Swarm Tactics Get Tryout for Infantry to Use in Urban Battlespace," *Army Times*, January 8, 2018, https://www.armytimes.com/news/your-army/2018/01/08/drone-swarm -tactics-get-tryout-for-infantry-to-use-in-urban-battlespace/; and Phillip Smith, "Drones and Swarm Behavior," *Drone Below*, July 10, 2018, https://dronebelow.com /drones-and-swarm-behavior.

2 T. Miller, *Storming the Wall* and Ahuja, *Planetary Specters*.

3 Since then drones provided intelligence in the Balkans in the 1990s, as well as reconnaissance and surveillance in Iraq and Afghanistan in the 1990s–2010s. For a striking account of the violence of US warfare and its visual mediation as humanitarian across wars from Vietnam to Afghanistan, see Atanasoski, *Humanitarian Violence*.

4 From DARPA'S website, https://www.darpa.mil/attachments/DARPAAccompli shmentsSeminalContributionstoNationalSecurity.pdf, accessed August 5, 2021. A separate section of DARPA's program is called Intelligence, Surveillance and Reconnaissance.

5 Whitney Webb, "Project Originally Funded by DARPA Seeks to Replace Bees with Tiny, Winged Robots," *True Activist*, October 6, 2016, https://www.trueactivist.com /project-originally-funded-by-darpa-seeks-to-replace-bees-with-tiny-winged-robots/.

6 Farriss, *Maya Society under Colonial Rule*.

7 Frost, *Biocultural Creatures*, 79.

8 Frost, *Biocultural Creatures*, 79.

9 Frost, *Biocultural Creatures*, 80.

10 For more on the secular/spiritual divide in feminist materialist scholarship, see Schaeffer, "Spirit Matters."

11 Scholars in postcolonial ecology and science studies question the isolation of "nature" as distinct from human intervention and the imperatives of empire and instead interrogate how insects and animals are altered through human interactions with the natural world. These scholars trespass the boundaries of segregated epistemologies in the academy that separate the natural sciences from human and social sciences, and nature from culture. For example, Jake Kosek argues that it is

not enough to ask what is happening to the bees to solve the problem of their disappearance, demanding that we look at the reengineering of human-bee environments, biology, and behaviors through US empire, or years of military use in battle,
intelligence, and testing. Kosek, "Ecologies of Empire." Also see the chapter "Can
the Mosquito Speak?," in Mitchell, *Rule of Experts,* 19-53; and Ahuja, *Bioinsecurities.*

12 Barad, *Meeting the Universe Halfway.*

13 Atanasoski and Vora, *Surrogate Humanity.*

14 Lewis et al., "Making Kin with the Machines." These scholars extend Donna
Haraway's idea of making kin discussed in her most recent book (and elsewhere),
Staying with the Trouble.

15 Lewis et al., "Making Kin with the Machines." They quote Vine Deloria Jr., who
wonders "why Western peoples believe they are so clever. Any damn fool can treat
a living thing as if it were a machine and establish conditions under which it is required
to perform certain functions—all that is required is a sufficient application of brute
force. The result of brute force is slavery." See V. Deloria, *Spirit and Reason,* 13.

16 Descola, *Beyond Nature and Culture,* 116.

17 Defense Advanced Research Projects Agency (DARPA), accessed July 14, 2020,
https://www.darpa.mil/news-events/darpa-redefining-possible.

18 The agency's website states, "The genesis of that mission and of DARPA itself dates
to the launch of Sputnik in 1957, and a commitment by the United States that, from
that time forward, it would be the initiator and not the victim of strategic technological surprises." "About DARPA," Defense Advanced Research Projects Agency,
accessed July 14, 2020, https://www.darpa.mil/about-us/about-darpa.

19 Arati Prabhakar (director of DARPA), statement, *Subcommittee on Intelligence, Emerging Threats and Capabilities, US House of Representatives* (March 26, 2014), 7, https://
www.darpa.mil/attachments/DrPrabhakar-26-Mar-14.pdf.

20 For an excellent discussion of the humanitarian justification of war for biomedical
developments, see Terry, *Attachments to War.*

21 See "Scientists Train Honeybees to Detect Explosives," Los Alamos National Lab,
Stealthy Insect Sensor Project, YouTube video, 5:59, posted March 21, 2008, https://
www.youtube.com/watch?v=_T7dobze4kM.

22 LaDuke, *Militarization of Indian Country.*

23 Quoted in Nick Turse, "DARPA's Wild Kingdom: Weaponized Bees, Robotic Rats,
Sleepless Soldiers; Does Mother Nature Stand a Chance in the Face of the Penta-
gon's New Science?," *Mother Jones,* March 8, 2004, https://www.motherjones.com
/politics/2004/03/darpas-wild-kingdom/.

24 See Terry, *Attachments to War,* for a closer look at the ways DARPA promises to
save and cure the very soldiers it commits to death, merging biopolitics with
necropolitics.

25 Numerous research articles and newspaper accounts confirm that the bee population in the United States declined from 5 million hives in 1950 to about 2.4 million
in 2006.

26 One of these researchers later moved this research to Harvard's microrobotics lab.
Webb, "Project Originally Funded by DARPA."

27 Amador and Hu, "Sticky Solution Provides Grip."

28 Kosek, "Ecologies of Empire."

29 In China workers pollinate crops such as apples with handheld pollen sticks. Apple trees, however, must be pollinated within five days of flowering, causing a severe challenge since one person can pollinate only five to ten trees per day. This would be too expensive in the United States and would cause even more outrage over the loss of bees. For more on this, see Stuart Liess, "After Bee Die-Off, Chinese Apple Farmers Resort to Hand Pollination," *Epoch Times*, April 25, 2015, http://www.theepochtimes.com/n3/1321746-after-bee-die-off-chinese-apple-farmers-resort-to-hand-pollination/.

30 For more information, see Radhika Nagpal, interview by Terrence McNally, on "Bioinspired Robotics," *Disruptive* (podcast), aired July 27, 2015, at 2:37, transcript available at http://aworldthatjustmightwork.com/2015/07/auto-draft-16/.

31 Nagpal, interview, at 8:11.

32 Dodge, *Hunting Grounds*, 368.

33 This discussion of labor can be found in Marx's, *Capital*, 174. For more discussion on Marx and bees, see Postone, *Time, Labor, and Social Domination*.

34 As argued by Ludwig Büchner, Descartes's assessment of animal consciousness borrows from a Spanish physician from the early 1600s who determined animals had no minds since they were machines controlled by external forces. Büchner, *Mind in Animals*.

35 The full title of Darwin's book is *On the Origin of Species by Means of Natural Selection, or the Preservation of Favoured Races in the Struggle for Life*. Darwin's scientific method is defined by his observations of animals from evidence he collected on the *Beagle* expedition in the 1830s, as well as research, experiments, and correspondence with other colleagues in a variety of fields, including zoologists, physicians, biologists, and other scientists.

36 Darwin, *On the Origin of Species*, 599.

37 Darwin, *On the Origin of Species*, 582.

38 Darwin argues that the hexagonal cells are stacked perfectly so that each cell fits perfectly with the side of another, thus lessening the amount of wax needed for each double-sided cell. And since a hive bee must produce more honey to produce more wax, efficient construction leads to less work. For more detail about how he worked with others on the mathematical solvency and accuracy of his argument, see S. Davis, "Darwin, Tegetmeier and the Bees."

39 Darwin, *On the Origin of Species*, 599. Darwin says that in their daily work, "they take what impoverishes none, while it enriches them and us also, by the valuable products which are derived from their skill and labour—true emblems of honest industry" (599).

40 Darwin refers to each bee group as a race rather than a species.

41 This book was originally written in German and then translated by Annie Besant into English.

42 Büchner, *Mind in Animals*, 266–67.

43 See Chavez, *Latino Threat*; and Tsing, "Empowering Nature."

44   *The Swarm* was directed and produced by Irwin Allen.

45   Nagpal, interview, at 8:11.

46   Brooks, *Cambrian Intelligence*, 138–39.

47   The influence of Grasse's term, *stigmagy*, is discussed in the first issues of the journal *Swarm Intelligence*. Garnier, Gautrais, and Theraulaz, "Biological Principles."

48   Nagpal, interview, at 30:30.

49   Nagpal, interview, at 27:55.

50   Andrew Quitmeyer and Tucker Balch, "The Waggle Dance of the Honeybee," Georgia Tech College of Computing, YouTube video, 7:28, posted February 2, 2011, https://www.youtube.com/watch?v=bFDGPgXtK-U.

51   See Villanueva-Gutiérrez, Roubik, and Colli-ucán, "Extinction of *Melipona beecheii*."

52   González-Acereto, Quezada-Euán, and Medina-Medina, "New Perspectives for Stingless Beekeeping."

53   This number may be a bit extreme since 2005 was a particularly severe year for hurricanes, which no doubt contributed to the decline in their numbers. Gwen Pearson, "Women Work to Save Native Bees of Mexico," *Wired*, March 5, 2014, https://www.wired.com/2014/03/women-work-save-native-bees-mexico/.

54   Quezada-Euán, De Jesús May-Itzá, and González-Acereto, "Meliponiculture in Mexico."

55   According to a study of beekeepers and the scholarship on Maya beekeeping, the *Melipona* "forage largely among secondary-growth plants." See Villanueva-Gutiérrez, Roubik, and Colli-ucán, "Extinction of *Melipona beecheii*."

56   Milpa agriculture is one of the most sophisticated planting methods and brings together the three sisters: corn, beans, and squash. Each plant contributes nutrients that aid the growth of the other and the grouping symbolizes the sacred number three as a crop that supports family, community, and the universe.

57   They succeeded in halting Monsanto from planting genetically modified soybeans in seven states across Mexico. These transgenic soybeans used by Monsanto (now owned by Bayer) are known as "Roundup Ready" due to their tolerance for high doses of the herbicide Roundup. Leydy Pech is convinced, as are many others, that these pesticides kill bees as they disorient the bees who lose their way home. And Monsanto's techniques for monocropping on a large scale wear down the soil and replace Maya farming centered on milpa planting and slash-and-burn farming, which has kept the land fertile for centuries. For her staunch defense of the environment, the bees, and the autonomy of the Maya, Pech won a 2020 Goldman Environmental Prize (also known as the "Green Nobel Prize").

58   Farriss, *Maya Society under Colonial Rule*.

59   See "Berta Cáceres: 2015 Goldman Prize Recipient, South and Central America," Goldman Environmental Prize, accessed July 15, 2019, https://www.goldmanprize .org/recipient/berta-caceres/.

60   Rob Nixon, "Indigenous Forest Defenders around the World Are Dying Anonymous Deaths," Literary Hub, January 16, 2020, https://lithub.com/indigenous-forest -defenders-around-the-world-are-dying-anonymous-deaths/. According to this article, a forest the size of Italy is cleared annually.

61 Quoted in the documentary "Xunan Kab—abeja del pueblo Maya," El Colegio de la Frontera Sur (with the help of historian Laura Sotelo and the Female Bee Collective, Koolel-Kab Cooperativa), YouTube video, 7:55 out of 27:07, posted May 19, 2016, https://www.youtube.com/watch?v=8QUjmr_U_yA.

62 For more on Anzaldúa's early analysis of her border crossing with the more-than-human, see Schaeffer, "Spirit Matters."

63 See Saldaña-Portillo, *Revolutionary Imagination in the Americas*, 282."

64 D. Perez, "New Tribalism and Chicana/o Indigeneity."

65 Anzaldúa, *Borderlands/La Frontera* and *Light in the Dark*.

66 Anzaldúa, "Yo, Gloria-2," n.d., Gloria Evangelina Anzaldúa Papers. Benson Latin American Collection, University of Texas Libraries, the University of Texas at Austin.

67 Aveni, *People and the Sky*, 192.

68 Aveni, *People and the Sky*, 192.

69 Gonzales, *Red Medicine*.

70 Anzaldúa, "En el Naranjal," box 70.7, n.d., Gloria Evangelina Anzaldúa Papers.

71 I am using terms coined by Gonzales in *Red Medicine*. Also see Zepeda, "Queer Xicana Indígena Cultural Production."

72 Rifkin, *Erotics of Sovereignty*, 73.

73 Tallbear, "Making Love and Relations."

74 In Shorter's essay "Sexuality," he suggests "that sexuality's power might be forceful enough to soothe the pains of colonization and the scars of internal colonization." See Shorter, "Sexuality." 487.

75 Tallbear, "The Critical Polyamorist," April 22, 2018, http://www.criticalpolyamorist .com/homeblog/yes-your-pleasure-yes-self-love-and-dont-forget-settler-sex-is-a -structure.

76 Shorter, "Spirituality," 446.

77 Shorter, "Sexuality," 497.

78 Lopez-Maldonado, "Ethnohistory of the Stingless Bees," 32.

79 De Jong, "Land of Corn and Honey," 57.

80 De Jong, "Land of Corn and Honey."

81 Sotelo Santos and Alvarez Asomoza, "Maya Universe."

82 M. Bianet Castellanos asks how rural to urban migration affects Maya communities in urban and rural areas across the Yucatán Peninsula and Quintana Roo in her book *A Return to Servitude: Maya Migration and the Tourist Trade in Cancún*. She finds that Maya migrants make up one-third of Cancún's population who work in the hotel industry, private homes, and construction (xviii).

83 Other Maya beekeepers similarly discuss the sensitivity of bees, such as the common knowledge that "if a beekeeper visits a cemetery, he must not visit his bees for three weeks, because he will carry the sadness of the cemetery with him, and the bees would slowly dwindle away." In addition, "when a beekeeper dies, the bees will leave unless the inheritor of the bees goes immediately to the hives to tell the bees of the death, and to assure them that he will care for them in the future; the bees need to know that there is someone who cares about them." See N. Weaver and E. Weaver, "Beekeeping with the Stingless Bee," 16.

84 De Jong, "Land of Corn and Honey."

85 De Jong, "Land of Corn and Honey."

86 De Jong, "Land of Corn and Honey," 313.

87 This statement and the epigraph are from the documentary "Xunan Kab—abeja del pueblo Maya," El Colegio de la Frontera Sur (with the help of historian Laura Sotelo and the Female Bee Collective, Koolel-Kab Cooperativa), YouTube video, 27:07, posted May 19, 2016, https://www.youtube.com/watch?v=8QUjmr_U_yA. This poem was written by Eleuteria Pech Mo and interpreted by Marieli Ávilla Chan.

88 "Soy Abeja Maya," Educampo, Mexico, YouTube video, 1:42:00, posted October 22, 2018, https://www.youtube.com/watch?v=DsBb6HEXoIw. Also, the term *ontoepistemology* was popularized by Karen Barad in their book *Meeting the Universe Halfway: Quantum Physics and the Entanglement of Matter and Meaning.*

89 For example, see Fregoso, "Pluriversal Declaration of Human Rights."

90 Laura Pérez also discusses this as a decolonial concept in *Chicana Art: The Politics of Spiritual and Aesthetic Altarities.*

91 Castellanos, *Return to Servitude*, 152.

92 For example, in an online video by Unanymous A.I., "What Is Swarm AI?" Louis Rosenberg says, "If a swarm of bees are smarter than one human brain, then a swarm of human brains should be able to create a super intelligence. This super intelligence won't be an *alien intelligence*, but a human super intelligence . . . just smarter. We should be able to solve the world's problems such as poverty, inequality, unsustainability . . . and do it with our human values, morals, sensibilities." Rosenberg uses swarm intelligence to develop a human swarm (rather than an alien, or mechanized, device) with the goal of using crowdsourcing to predict things like the winner of a horse race and human consumption patterns. See Louis Rosenberg, "What Is Swarm AI?," YouTube video, 10:59, posted November 3, 2017, https://www.youtube.com/watch?v=xWSkbsIRNMg.

93 Lenkersdorf, "Tojolabal Language."

94 Lenkersdorf, "Tojolabal Language," 105.

CONCLUSION. WILD VERSUS SACRED

Epigraph: "BorderViews: Nellie Jo David, Hia-Ced O'odham Activist," Center for Biological Diversity, November 19, 2019, 1:38 minutes, https://www.youtube.com/watch?v=SShcWK1BhlY.

1 By the time President Trump left office, contractors had built over 450 miles of the new border wall. Close to 230 miles of it was erected in Arizona, according to the US Customs and Border Protection. Even though the new wall remains sixty miles from the border located in the Tohono O'odham Nation, it cuts into the ancestral land of the O'odham.

2 The US Army Corps of Engineers and Customs and Border Protection blew up an O'odham burial site on Monument Hill to clear land for the construction of the border wall. See "Tohono O'odham Chair Decries Ongoing Controlled Blasts for Border Wall Project," February 28, 2020, *Arizona Public Media*, https://www.youtube.com/watch?v=aobcQaGMRL8.

3 Different tribes of the O'odham had been at odds with various bands of neighboring Apache from the seventeenth century to the twentieth century. These conflicts came to a head in 1871 at the Camp Grant Massacre when O'odham joined Mexicans and Anglos and killed 144 Apache, many of them women and children. Over time, both have incorporated enemies into their kin structures. Paulina Firozi, "Sacred Native American Burial Sites Are Being Blown Up for Trump's Border Wall, Lawmaker Says," *Washington Post*, February 9, 2020, https://www.washingtonpost .com/immigration/2020/02/09/border-wall-native-american-burial-sites/.

4 Many Indigenous peoples, like many others around the world, know that until the dead are laid to rest, their people will suffer social, psychic, economic, and bodily illness. For one example, see Yellowman, "'Naevahooohtseme,'" quoted in Kosslak, "Native American Graves," 134.

5 Quoted in Brenda Norrell, "Border Wall: United States Government Destroying O'odham Burial Place," *Censored News: Indigenous Peoples and Human Rights*, January 26, 2020, https://bsnorrell.blogspot.com/2020/01/border-wall-united-states -government.html.

6 Flesch et al., "Potential Effects," 172.

7 Sarah Jaquette Ray has shown that a "green anti-immigrant rhetoric" denies the ways the US-Mexico border landscape was first cleared of Indigenous and then later Mexican populations to become "wilderness." See Ray, *Ecological Other*, 14–15.

8 The DOI manages about 75 percent of federal land. They are charged with land and natural resource management, American Indian affairs, wildlife conservation, and territorial affairs.

9 Gerald S. Dickinson, "Forget Funding the Wall, Trump Needs the Land First," *Hill*, August 25, 2017, https://thehill.com/blogs/pundits-blog/the-administration/347912 -forget-funding-the-wall-trump-needs-the-land-first.

10 *Destroying Sacred Sites and Erasing Tribal Culture: The Trump Administration's Construction of the Border Wall; Oversight Hearing before the Subcommittee for Indigenous Peoples of the United States of the Committee of Natural Resources, U.S. House of Representatives*, 116th Congress (February 26, 2020), https://www.congress.gov/event/116th-congress /house-event/LC65186/text?s=1&r=15.

11 *Destroying Sacred Sites*, 70.

12 Roosevelt declared this borderland region in California, Arizona, and New Mexico. Even when the federal government claimed the land, wealthy ranchers continued to graze cattle on these lands for over a decade, destroying the O'odham ecological and cultural practices on this land.

13 Greene, "Historic Resource Study," 62.

14 See Riding In, "Six Pawnee Crania"; and Trope and Echo-Hawk, "Native American Graves," 42.

15 For more on this, see Urbanski, "Defend the Sacred."

16 Lonetree, *Decolonizing Museums*.

17 *Impact of the Drug Trade* (April 15, 2003) (statement of Frank Deckert, Superintendent, Big Bend National Park, National Park Service, Department of the Interior), 19.

18 Blanchfield and Kolowratnik, "At the Border."

19   *Destroying Sacred Sites video* (statement of Paul A. Gosar, DOI, 1:46), https://www
     .congress.gov/event/116th-congress/house-event/110587.

20   This DOI perspective is well supported by a long history of environmental racism
     against immigrants. At the turn of the twentieth century, environmental preserva-
     tionists blamed immigrants from eastern Europe, Italy, and China and even Black
     migrants for the rise in urban pollution and the destruction of wilderness areas. See
     Sadowski-Smith, "U.S. Border Ecologies"; and Park and Pellow, "Roots of Nativist
     Environmentalism."

21   Section 102 of the Real ID Act states, "The Secretary of Homeland Security shall
     have the authority to waive, and shall waive, all laws such Secretary, in such Secre-
     tary's sole discretion, determines necessary to ensure expeditious construction of
     the barriers and roads under this Section." See the Real ID act of 2005, 8 U.S.C. §
     1101 (2008).

22   In 1923, President Calvin Coolidge stole forty acres of land around Quitobaquito
     Springs and designated the area Public Water Reserve No. 88. See Green, "Historic
     Resource Study," 61.

23   See the Sierra Club's Borderland Campaign video "Wild Versus Wall." Steev Hise,
     the Border Campaign of the Sierra Club, Grand Canyon Chapter, Arizona (2010),
     https://vimeo.com/9561480.

24   Candice Bernd, "Biden Is Rejecting Trump's Border Wall—but Favors His Own
     Technological Wall," *Truthout*, February 2, 2021, https://truthout.org/articles/biden
     -is-rejecting-trumps-border-wall-but-proposing-his-own-virtual-wall.

25   Fojas, *Border Optics*.

26   *Destroying Sacred Sites* (statement of Scott J. Cameron, principal deputy assistant
     secretary for policy, management, and budget, US DOI), 70.

27   See Tim Tibbitts and Mark Sturm, "Wild Matters: The Organ Pipe Cactus
     Wilderness," National Park Service, December 2011, https://www.nps.gov/orpi
     /planyourvisit/upload/Wild-Matters_The-Organ-Pipe-Cactus-Wilderness_2020
     .pdf.

28   National Parks Conservation Association, "President Trump's Proposed Budget
     Cuts Target National Parks," February 10, 2020, https://www.npca.org/articles/2457
     -president-trump-s-proposed-budget-cuts-target-national-parks.

29   Division of Natural and Cultural Resources Management, *Organ Pipe Cactus Na-
     tional Monument, Ecological Monitoring Program, Annual Report 1993*," 1.

30   As "a pristine example of an intact Sonoran Desert ecosystem," Organ Pipe was
     designated as a UNESCO International Biosphere Reserve in 1976.

31   Organ Pipe Cactus National Monument, "Organ Pipe Cactus National Monument:
     Superintendent's 2010 Report on Natural Resource Vital Signs," https://www.nps
     .gov/orpi/learn/nature/upload/orpi_vitalsigns2010.pdf, 4.

32   *Impact of the Drug Trade* (April 15, 2003).

33   See Stern, *Eugenic Nation*; Solnit, *Savage Dreams*; Cronon, "Trouble with Wilderness."

34   O'Brien, *Firsting and Lasting*; M. K. Anderson, *Tending the Wild*.

35   During the Black Lives Matter protests, when statues of racist national figures
     were targeted in attempts to take back the city of New York from racism and police

violence, the New York governor ultimately agreed that it was time to take down Teddy Roosevelt's statue. See Robin Pogrebin, "Roosevelt Statue to Be Removed from Museum of Natural History," *New York Times*, June 21, 2020, https://www.nytimes.com/2020/06/21/arts/design/roosevelt-statue-to-be-removed-from-museum-of-natural-history.html.

36 Wilderness Act of 1964, 16 U.S.C. 1131, available at https://www.nps.gov/orgs/1981/upload/W-Act_508.pdf.

37 M. K. Anderson, *Tending the Wild*.

38 Norrell, "Border Wall."

39 *Destroying Sacred Sites and Erasing Tribal Culture*, February 26, 2020, https://www.govinfo.gov/content/pkg/CHRG-116hhrg40260/html/CHRG-116hhrg40260.htm.

40 To learn more about the use of footprints on precolonial maps used by the Maya, see Hidalgo, *Trail of Footprints*.

41 Ofelia Rivas (elder and activist), "Ofelia Rivas, the Tohono O'odham, and the Wall," interview by Marcello Di Cintio, *Elsewhere*, January 19, 2019, https://marcellodicintio.com/2019/01/19/ofelia-rivas-the-tohono-oodham-and-the-wall/.

42 Had Anzaldúa lived on, I am sure she would have come to this understanding of her work.

43 In his recent book, Todd Miller equates militarized surveillance at the US border with the title of his book, *Empire of Borders*.

44 Anzaldúa, *Borderlands*, 24–25.

Adey, Peter. "Facing Airport Security: Affect, Biopolitics, and the Preemptive Securitization of the Mobile Body." *Environmental and Planning D: Society and Space* 27, no. 3 (2009): 274–95.

Ahmed, Sara. *Queer Phenomenology: Orientations, Objects, Others*. Durham, NC: Duke University Press, 2006.

Ahuja, Neel. *Bioinsecurities: Disease Interventions, Empire, and the Government of Species*. Durham, NC: Duke University Press, 2016.

Ahuja, Neel. *Planetary Specters: Race, Migration and Climate Change in the Twenty-First Century*. Chapel Hill: University of North Carolina Press, 2021.

Ajana, Btihaj. *Governing through Biometrics: The Biopolitics of Identity*. Hampshire, UK: Palgrave Macmillan, 2013.

Ajana, Btihaj. "Recombinant Identities: Biometrics and Narrative Bioethics." *Journal of Bioethical Inquiry* 7, no. 2 (2010): 237–58.

Alberto, Lourdes. "Mexican American Indigeneities." New York: New York University Press, forthcoming.

Alexander, M. Jacqui. *Pedagogies of Crossing: Meditations on Feminism, Sexual Politics, Memory, and the Sacred*. Durham, NC: Duke University Press, 2005.

Allen, Irwin, dir. *The Swarm*. Warner Bros., 1978. 1 hr., 56 min. https://www.amazon.com/gp/video/detail/amzn1.dv.gti.18a9f74a-a89d-3190-4da0-0563103f8894?autoplay=1&ref_=atv_cf_strg_wb.

Amador, Guillermo J., and David L. Hu. "Sticky Solution Provides Grip for the First Robotic Pollinator." *Chem* 2, no. 2 (2017): 162–64.

Amoore, Louise. "Biometric Borders: Governing Mobilities in the War on Terror." *Political Geography* 25, no. 3 (2006): 336–51.

Anderson, Bridget. "More Equal Than Others." In *Precarity and Citizenship*, edited by Sylvanna Falcón, Steve McKay, Juan Poblete, Catherine S. Ramírez, and Felicity Amaya Schaeffer, 19–31. New Brunswick, NJ: Rutgers University Press, 2021.

Anderson, M. Kat. *Tending the Wild: Native American Knowledge and the Management of California's Natural Resources*. Berkeley: University of California Press, 2005.

Andreas, Peter. *Border Games: Policing the U.S.-Mexico Divide*. Ithaca, NY: Cornell University Press, 2009.

Andrejevic, Marc. *Infoglut: How Too Much Information Is Changing the Way We Think and Know*. New York: Routledge, 2013.

Anzaldúa, Gloria. *Borderlands/La Frontera: The New Mestiza*. San Francisco: Aunt Lute Books, 1987.

Anzaldúa, Gloria. "En el Naranjal," box 70.7, n.d. Gloria Evangelina Anzaldúa Papers. Benson Latin American Collection, Austin: University of Texas Libraries.

Anzaldúa, Gloria. *Light in the Dark/Luz en lo Oscuro*. Edited by Analouise Keating. Durham, NC: Duke University Press, 2015.

Anzaldúa, Gloria. "Yo, Gloria-2," n.d. Gloria Evangelina Anzaldúa Papers. Benson Latin American Collection, Austin: University of Texas Libraries.

Arizona Department of Health Services. *Arizona Health Status and Vital Statistics 2005 Report*. Phoenix: Arizona Department of Health Services, 2006. https://pub.azdhs.gov/health-stats/report/ahs/ahs2005/index.htm.

Army Intelligence Museum, Fort Huachuca. "The Intelligence Impulse: A Showcase for U.S. Army Intelligence History." Accessed August 3, 2021. https://docplayer.net/82910164-The-intelligence-impulse-a-showcase-for-u-s-army-intelligence-history.html.

Atanasoski, Neda. *Humanitarian Violence: The U.S. Deployment of Diversity*. Minneapolis: University of Minnesota Press, 2013.

Atanasoski, Neda, and Kalindi Vora. *Surrogate Humanity: Race, Robots, and the Politics of Technological Futures*. Durham, NC: Duke University Press, 2019.

Aveni, Anthony. *People and the Sky: Our Ancestors and the Cosmos*. New York: Thames and Hudson, 2008.

Ball, Eve. *In the Days of Victorio: Recollections of a Warm Springs Apache*. Tucson: University of Arizona Press, 2008.

Barad, Karen. *Meeting the Universe Halfway: Quantum Physics and the Entanglement of Matter and Meaning*. Durham, NC: Duke University Press, 2007.

Barad, Karen. "Troubling Time/s and Ecologies of Nothingness: Re-turning, Re-membering, and Facing the Incalculable." *New Formations* 92 (2017): 56–86.

Barker, Joanne. *Native Acts: Law, Recognition, and Cultural Authenticity*. Durham, NC: Duke University Press, 2011.

Barker, Joanne. "Territory as Analytic: The Dispossession of Lenapehoking and the Subprime Crisis." *Social Text* 36, no. 2/135 (2018): 19–39.

Barr, Juliana. *Peace Came in the Form of a Woman: Indians and Spaniards in the Texas Borderlands*. Chapel Hill: University of North Carolina Press, 2007.

Barrett, Stephen M., ed. and trans. *Geronimo's Story of His Life*. New York: Duffield, 1906. Kindle.

Basso, Keith H. *Wisdom Sits in Places: Landscape and Language among the Western Apache*. Albuquerque: University of New Mexico Press, 1996.

Baynton, Douglas C. *Forbidden Signs: American Culture and the Campaign against Sign Language*. Chicago: University of Chicago Press, 1998.

Benjamin, Ruha. "Catching Our Breath: Critical Race STS and the Carceral Imagination." *Engaging Science, Technology, and Society* 2 (2016): 145–56.

Benjamin, Ruha. "Introduction: Discriminatory Design, Liberating Imagination." In *Captivating Technology: Race, Carceral Technoscience, and Liberatory Imagination in Everyday Life*, edited by Ruha Benjamin, 1–22. Durham, NC: Duke University Press, 2019.

Benton, Lauren. "Spatial Histories of Empire." *Itinerario* 30, no. 3 (2006): 19–34.

Benton, Lauren, and Benjamin Straumann. "Acquiring Empire by Law: From Roman Doctrine to Early Modern European Practice." *Law and History Review* 28, no. 1 (2010): 1–38.

Berlant, Lauren. "Slow Death (Sovereignty, Obesity, Lateral Agency)." *Critical Inquiry* 33, no. 4 (Summer 2007): 754–80.

Bianchini, Marcello Levi. "La mentalità della razza calabrese: Saggio di psicologia etnica." *Rivista Psicologica Applicata alla Pedagogia ed alla Psicopatologia* 2, no. 1 (1906): 13–21.

Blackwell, Maylei, Floridalma Boj Lopez, and Luis Urrieta Jr. "Critical Latinx Indigeneities." Special issue, *Latino Studies* 15, no. 2 (2017).

Blackwell, Maylei, Floridalma Boj Lopez, and Luis Urrieta Jr., eds. Introduction to "Critical Latinx Indigeneities." Special issue, *Latino Studies* 15, no. 2 (2017): 126–37.

Blanchfield, Caitlin, and Nina Valerie Kolowratnik. "'Persistent Surveillance': Militarized Infrastructure on the Tohono O'odham Nation." *Avery Review* 40 (May 2019). https://www.averyreview.com/issues/40/persistent-surveillance.

Bolkcom, Christopher. *Homeland Security: Unmanned Aerial Vehicles and Border Surveillance.* Congressional Research Service Report for Congress, June 28, 2004, Washington, DC. https://fas.org/sgp/crs/RS21698.pdf.

Bonds, Anne, and Joshua Inwood. "Beyond White Privilege: Geographies of White Supremacy and Settler Colonialism." *Progress in Human Geography* 40, no. 6 (2016): 715–33.

Bourke, John G. *An Apache Campaign in the Sierra Madre.* New York: Charles Scribner's Sons, 1886.

Brandt, Elizabeth A. "The Fight for Dzil Nchaa Si An, Mt. Graham." *Cultural Survival Quarterly* 19, no. 4 (Winter 1996): 50–57.

Brooks, Rodney A. *Cambrian Intelligence: The Early History of the New AI.* Cambridge: MA, MIT Press, 1999.

Brown, Wendy. *Walled States, Waning Sovereignty.* Brooklyn, NY: Zone, 2010.

Browne, Simone. *Dark Matters: On the Surveillance of Blackness.* Durham, NC: Duke University Press, 2015.

Browne, Simone. "Digital Epidermalization: Race, Identity, Biometrics." *Critical Sociology Journal*, 36, no. 1 (February 2010): 131–50.

Bruyneel, Kevin. *The Third Space of Sovereignty: The Postcolonial Politics of U.S.-Indigenous Relations.* Minneapolis: University of Minnesota Press, 2007.

Büchner, Ludwig, *Mind in Animals.* Translated by Annie Besant. London: Freethought, 1880.

Burgoon, Judee K., Laura K. Guerrero, and Valerie Mansov. "Nonverbal Signals." In *The Sage Handbook of Interpersonal Communication*, edited by Mark L. Knapp and John A. Daly, 239–80. 4th ed. Thousand Oaks, CA: Sage, 2011.

Burgoon, Judee K., Jeffrey G. Proudfoot, Ryan Schuetzler, and David Wilson. "Patterns of Nonverbal Behavior Associated with Truth and Deception: Illustrations from Three Experiments." *Journal of Nonverbal Behavior* 38, no. 3 (September 2014): 325–54.

Burns, Robert P. "Privacy Impact Assessment for the Future Attribute Screening Technology (FAST) Project." Department of Homeland Security, December 15, 2008, Washington, DC. https://www.dhs.gov/xlibrary/assets/ privacy/privacy_pia_st_fast.pdf.

Butler, Judith. *Bodies that Matter: On the Discursive Limits of "Sex."* New York: Routledge, 1993.

Byrd, Jodi. *Transit of Empire: Indigenous Critiques of Colonialism.* Minneapolis: University of Minnesota Press, 2011.

Cadava, Geraldo L. "Borderlands of Modernity and Abandonment: The Lines within Ambos Nogales and the Tohono O'odham Nation." *Journal of American History* 98, no. 2 (September 2011): 362–83. https://doi.org/10.1093/jahist/jar209.

Cajete, Gregory. *Native Science: Natural Laws of Interdependence.* Santa Fe, NM: Clear Light Books, 2000.

Carte, Gene E., and Elaine H. Carte. *Police Reform in the United States: The Era of August Vollmer, 1905–1932.* Berkeley: University of California Press, 1975.

Castellanos, M. Bianet. *A Return to Servitude: Maya Migration and the Tourist Trade in Cancún.* Minneapolis: University of Minnesota Press, 2010.

Castellanos, M. Bianet, Lourdes Gutiérrez Nájera, and Arturo J. Aldama, eds. *Comparative Indigeneities of the Américas: Towards a Hemispheric Approach.* Tucson: University of Arizona Press, 2012.

Castillo, Ana. *Massacre of the Dreamers: Essays on Xicanisma.* New York: Plume Books, 1995.

Chacón, Justin Akers, and Mike Davis. *No One Is Illegal: Fighting Violence and the State Repression on the U.S.-Mexico Border.* Chicago: Haymarket Books, 2006.

Chavez, Leo R. *The Latino Threat: Constructing Immigrants, Citizens, and the Nation.* Palo Alto, CA: Stanford University Press, 2013.

Chow, Rey. *The Age of the World Target: Self-Referentiality in War, Theory and Comparative Work.* Durham, NC: Duke University Press, 2006.

Clark, W. P. *The Indian Sign Language.* Lincoln: University of Nebraska Press, 1982.

Cole, Simon A. *Suspect Identities: A History of Fingerprinting and Criminal Identification.* Cambridge, MA: Harvard University Press, 2001.

Cortés y de Olarte, José María. *Views from the Apache Frontier: Report on the Northern Provinces of New Spain (Memoria sobre las provincias del norte de Nueva Espana).* Edited by Elizabeth A. H. John. Translated by John Wheat. Norman: University of Oklahoma Press, 1994 [Spain, 1799].

Cronon, William. "The Trouble with Wilderness; or, Getting Back to the Wrong Nature." In *Uncommon Ground: Rethinking the Human Place in Nature,* edited by William Cronon, 69–90. New York: W. W. Norton, 1996.

Cruz, Robert. "AM T.E'OK ET A:T O CE:EKT DO'IBIODA:LIK"/"In Our Language Is Where We Will Find Our Liberation." *Berkeley La Raza Law Journal* 22, no. 5 (2012): 97–115.

Cuarón, Jonás, dir. *Desierto.* Burbank, CA: STXfilms, 2015. 1 hr., 27 min. https://tubitv.com/movies/607432/desierto?start=true&utm_source=google-feed&tracking=google-feed.

Cuvier, Georges. *Lectures on Comparative Anatomy.* London: Wilson Fort, N. Longman, and O. Rees, 1802.

Darwin, Charles. *The Descent of Man.* 1879. London: Penguin Books, 2004.

Darwin, Charles. *The Expression of the Emotions in Man and Animals.* 1872. Oxford: Oxford University Press, 1998.

Darwin, Charles. *On the Origin of Species by Means of Natural Selection, or the Preservation of Favoured Races in the Struggle for Life.* London: John Murray, 1859.

Davis, Jeffrey E. *Hand Talk: Sign Language among American Indian Nations.* Cambridge: Cambridge University Press, 2010.

Davis, Sarah. "Darwin, Tegetmeier and the Bees." *Science Part C: Studies in History and Philosophy of Biological and Biomedical Sciences* 35, no. 1 (March 2004): 65–92.

De Galvez, Bernando. *Instructions for Governing the Interior Provinces of New Spain.* Berkeley: Quivera Society, 1787.

De Genova, Nicholas. "Denizenship." In *Precarity and Citizenship*, edited by Sylvanna Falcón, Steve McKay, Juan Poblete, Catherine S. Ramírez, and Felicity Amaya Schaeffer, 227–42. New Brunswick, NJ: Rutgers University Press, 2021.

De Jong, H. "The Land of Corn and Honey: The Keeping of Stingless Bees (Meliponiculture) in the Ethno-Ecological Environment of Yucatán, Mexico and El Salvador." PhD diss., Utrecht University, 1999.

De León, Jason. *The Land of Open Graves: Living and Dying on the Migrant Trail.* Berkeley: University of California Press, 2015.

Deloria, Philip. *Indians in Unexpected Places.* Lawrence: University Press of Kansas, 2004.

Deloria, Philip. *Playing Indian.* New Haven, CT: Yale University Press, 1998.

Deloria, Vine, Jr. *Spirit and Reason: The Vine Deloria Jr. Reader.* Edited by Barbara Deloria, Kristen Foehner, and Sam Scinta. Golden, CO: Fulcrum, 1999.

Descola, Philippe. *Beyond Nature and Culture.* Translated by Janet Lloyd. Chicago: University of Chicago Press, 2013.

Díaz-Barriga, Miguel, and Margaret E. Dorsey, eds. *Fencing In Democracy: Border Walls, Necrocitizenship and the Security State.* Durham, NC: Duke University Press, 2020.

Dijstelbloem, Huub, Albert Meijer, and Michiel Besters. "The Migration Machine." In *Migration and the New Technological Borders of Europe: Migration, Minorities and Citizenship*, edited by Huub Dijstelbloem and Albert Meijer, 1–21. Basingstoke, UK: Palgrave Macmillan, 2011.

Dillon, Grace. Introduction to *Walking the Clouds: An Anthology of Indigenous Science Fiction*, edited by Grace L. Dillon, 1–12. Tucson: University of Arizona Press, 2012.

Division of Natural and Cultural Resources Management, *Organ Pipe Cactus National Monument, Ecological Monitoring Program, Annual Report 1993.* Tucson, AZ: National Biological Service, Cooperative Park Studies Unit, University of Arizona, 1995.

Dodge, Richard Irving. *The Hunting Grounds of the Great West: A Description of the Plains, Game, and Indians of the Great North American Desert.* London: Chatto and Windus, 1877.

Dunbar-Ortiz, Roxanne. *An Indigenous People's History of the United States.* Boston: Beacon, 2014.

Dunn, Timothy. *The Militarization of the U.S.-Mexico Border, 1978–1992: Low Intensity Conflict Doctrine Comes Home.* Austin: University of Texas Press, 1996.

Ekman, Paul. "Facial Expression and Emotion." *American Psychologist* 48, no. 4 (1993): 384–92.

Ekman, Paul. *Telling Lies: Clues to Deceit in the Marketplace, Politics, and Marriage.* New York: W. W. Norton, 2001.

Ekman, Paul, and Wallace V. Friesen. *Facial Action Coding System: A Technique for the Measurement of Facial Movement.* Palo Alto, CA: Consulting Psychologists Press, 1978.

Ekman, Paul, and Wallace V. Friesen. "Nonverbal Leakage and Clues to Deception." *Psychiatry* 32, no. 1 (1969): 88–105.

Ekman, Paul, Wallace V. Friesen, and Klaus Scherer. "Body Movement and Voice Pitch in Deceptive Interaction." *Semiotica* 16, no. 1 (1979): 23–27.

Escobar, Arturo. *Designs for the Pluriverse: Radical Interdependence, Autonomy, and the Making of Worlds*. Durham, NC: Duke University Press, 2018.

Escobar, Arturo. "Thinking-Feeling with the Earth: Territorial Struggles and the Ontological Dimension of the Epistemologies of the South." *Revista de Antropología Iberoamericana* 11, no. 1 (2016): 11–32.

Facio, Elisa, and Irene Lara, eds. *Fleshing the Spirit: Spirituality and Activism in Chicana, Latina, and Indigenous Women's Lives*. Tucson: University of Arizona Press, 2014.

Fanon, Frantz. *White Skin, Black Masks*. Translated by Richard Philcox. 1952. New York: Grove, 2008.

Farriss, Nancy M. *Maya Society under Colonial Rule: The Collective Enterprise of Survival*. Princeton, NJ: Princeton University Press, 1984.

Fassin, Didier. "Compassion and Repression: The Moral Economy of Immigration Policies in France." *Cultural Anthropology* 20, no. 3 (2005): 362–87.

Fernandes, Sujatha. *Curated Stories: The Uses and Misuses of Storytelling*. New York: Oxford University Press, 2017.

Finley, James P. *The Buffalo Soldiers at Fort Huachuca*. Vols. 1–3 of *Huachuca Illustrated: A Magazine of the Fort Huachuca Museum*. Sierra Vista, AZ: Huachuca Museum Society, 1993–96.

Flesch, Aaron D., Clinton W. Epps, James W. Cain III, and Matt Clark. "Potential Effects of the United States-Mexico Border Fence on Wildlife." *Conservation Biology* 24, no. 1 (2009): 171–81.

Flynn, Michael T., Rich Juergens, and Thomas L. Cantrell. "Employing ISR: Special Operations Forces (SOF) Best Practices." *Joint Force Quarterly*, no. 50 (2008): 56–61.

Fojas, Camilla. *Border Optics: Surveillance on the U.S.-Mexico Frontier*. New York: New York University Press, 2021.

Forbes, Gordon, III. "Shadow Wolves: ICE'S Native American Manhunters." Des Plaines, IL: Topics Media Group, 2009.

Foucault, Michel. *The History of Sexuality*. Translated by Robert Hurley. New York: Pantheon Books, 1978.

Foucault, Michel. "*Society Must Be Defended: Lectures at the Collège de France, 1975–76*. Translated by David Macy. New York: Picador, 2003.

Fregoso, Rosa-Linda. "For a Pluriversal Declaration of Human Rights." In "Las Americas Quarterly." Special issue, *American Quarterly* 66, no. 3 (September 2014): 583–608.

Fregoso, Rosa-Linda. *MeXicana Encounters: The Making of Social Identities on the Borderlands*. Berkeley: University of California Press, 2003.

Freud, Sigmund. *Totem and Taboo: Some Points of Agreement between the Mental Lives of Savages and Neurotics*. Translated by James Strachey, London: Routledge, 2001.

Frost, Samantha. *Biocultural Creatures: Toward a New Theory of the Human*. Durham, NC: Duke University Press, 2016.

Ganster, Paul. *The U.S.-Mexican Border Today: Conflict and Cooperation in Historical Perspective*. With David E. Lorey. Lanham, MD: Rowman and Littlefield, 2016.

Garnier, Simon, Jaques Gautrais, and Guy Theraulaz. "The Biological Principles of Swarm Intelligence." *Swarm Intelligence* 1 (2007): 3–31.

Gates, Kelly. *Our Biometric Future: Facial Recognition Technology and the Culture of Surveillance*. New York: New York University Press, 2011.

Goeman, Mishuana. "Land as Life: Unsettling the Logics of Containment." In *Native Studies Keywords*, edited by Stephanie Nohelani Teves, Andrea Smith, and Michelle H. Raheja, 71–89. Tucson: University of Arizona Press, 2015.

Goeman, Mishuana. *Mark My Words: Native Women Mapping Our Nations*. Minneapolis: University of Minnesota Press, 2013.

Golash-Boza, Tanya. "The Parallels between Mass Incarceration and Mass Deportation: An Intersectional Analysis of State Repression." *Journal of World-Systems Research* 22, no. 2 (2016): 484–509.

Gómez-Barris, Macarena. *The Extractive Zone: Social Ecologies and Decolonial Perspectives*. Durham, NC: Duke University Press, 2017.

Gonzales, Patrisia. *Red Medicine: Traditional Indigenous Rites of Birthing and Healing*. Tucson: University of Arizona Press, 2012.

González-Acereto, J. A., Jose Javier G. Quezada-Euán, and Luis Medina-Medina. "New Perspectives for Stingless Beekeeping in the Yucatan: Results of an Integral Program to Rescue and Promote the Activity." *Journal of Apicultural Research* 45, no. 4 (2006): 234–39.

Goodwin, Neil, dir. *Geronimo and the Apache Resistance*. Cambridge, MA: Peace River Films, 1988. DVD, 56 min.

Gordillo, Gastón. "Terrain as Insurgent Weapon: An Affective Geometry of Warfare in the Mountains of Afghanistan." *Political Geography* 64 (May 2018): 53–62.

Green, Lesley, and David R. Green. *Knowing the Day, Knowing the World: Engaging Amerindian Thought in Public Archaeology*. Tucson: University of Arizona Press, 2013.

Greene, Jerome A. "Historic Resource Study, Organ Pipe Cactus National Monument, Arizona." Report for the Historic Preservation Division, National Park Service, Department of the Interior, Denver, CO, September 1977.

Gregory, Derek. "Dirty Dancing: Drones and Death in the Borderlands." In *Life in the Age of Drone Warfare*, edited by Lisa Parks and Caren Kaplan, 25–58. Durham, NC: Duke University Press, 2017.

Guidotti-Hernández, Nicole. *Unspeakable Violence: Remapping U.S. and Mexican National Imaginaries*. Durham, NC: Duke University Press, 2011.

Gündogdu, Ayten. *Rightlessness in an Age of Rights: Hannah Arendt and the Contemporary Struggles of Migrants*. Oxford: Oxford University Press, 2015.

Haney, William M. "Protecting Tribal Skies: Why Indian Tribes Possess the Sovereign Authority to Regulate Tribal Airspace." *American Indian Law Review* 4, no. 1 (2016): 1–40.

Haraway, Donna J. *Staying with the Trouble: Making Kin in the Chthulucene*. Durham, NC: Duke University Press, 2016.

Hernandez, Kelly Lytle. *City of Inmates: Conquest, Rebellion, and the Rise of Human Caging in Los Angeles 1771–1965*. Chapel Hill: University of North Carolina Press, 2020.

Hernandez, Kelly Lytle. *Migra! A History of the U.S. Border Patrol*. Berkeley: University of California Press, 2010.

Hidalgo, Alex. *Trail of Footprints: A History of Indigenous Maps from Viceregal Mexico*. Austin: University of Texas Press, 2019.

Higginbotham, Adam. "Deception is Futile: In Search of the Perfect Lie Detector Test." *Wired*, February 2013. http://www.adamhigginbotham.com/Archive/Writing_files/2102FF_deception_spreads.pdf.

Hise, Steev, prod. "Wild Versus Wall." Tucson, AZ: Pan Left Productions, 2010 (revised). 19:10 minutes. https://vimeo.com/9561480.

Hobart, Hi'ilei Julia. "At Home on the Mauna: Ecological Violence and Fantasies of Terra Nullius on Maunakea's Summit." *Native American and Indigenous Studies* 6, no. 2 (Fall 2019): 30–50.

Hochman, Brian. *Savage Preservation: The Ethnographic Origins of Modern Media Technology*. Minneapolis: University of Minnesota Press, 2014.

Hu, Tung-Hui. *A Prehistory of the Cloud*. Cambridge, MA: MIT Press, 2015.

Hubbard, Tasha. "Buffalo Genocide in Nineteenth-Century North America: 'Kill, Skin, and Sell.'" In *Colonial Genocide in Indigenous North America*, edited by Alexander Laban Hinton, Andrew Woolford, and Jeff Benvenuto, 292–305. Durham, NC: Duke University Press, 2014.

Human Rights Watch. *Forced Apart (by the Numbers): Non-citizens Deported Mostly for Nonviolent Offences*. New York: Human Rights Watch, 2009.

Kao, John. *Innovation Nation: How America Is Losing Its Innovation Edge, Why It Matters, and What We Can Do to Get It Back*. New York: Free Press, 2007.

Kaplan, Caren. *Aerial Aftermaths: Wartime from Above*. Durham, NC: Duke University Press, 2018.

Katzenbach, Christina. "Algorithmic Governance." *Internet Policy Review: Journal on Internet Regulation* 8, no. 4 (2019): 1–18.

Katzenbach, Christian, and Lena Ulbricht. 2019. "Algorithmic governance." *Internet Policy Review* 8, no. 4 (November 29, 2019). https://policyreview.info/concepts/algorithmic-governance.

Kauanui, J. Kēhaulani. *Hawaiian Blood: Colonialism and the Politics of Sovereignty and Indigeneity*. Durham, NC: Duke University Press, 2008.

Kimmerer, Robin. "Weaving Traditional Ecological Knowledge into Biological Education: A Call to Action." *Bioscience* 52, no. 5 (2002): 432–38.

Klein, Naomi. *The Shock Doctrine: The Rise of Disaster Capitalism*. New York: Picador, 2007.

Kosek, Jake. "Ecologies of Empire: On the New Uses of the Honeybee." *Cultural Anthropology* 25, no. 4 (2010): 650–78.

Kosslak, Renee M. "The Native American Graves Protection and Repatriation Act: The Death Knell for Scientific Study?" *American Indian Law Review* 24, no. 1 (1999): 129–51.

LaDuke, Winona. *The Militarization of Indian Country*. With Sean Aaron Cruz. East Lansing, MI: Makwa Enewed, 2013.

LaDuke, Winona. *Recovering the Sacred: The Power of Naming and Claiming*. Cambridge, MA: South End, 2005.

Laluk, Nicholas C. "The Indivisibility of Land and Mind: Indigenous Knowledge and Collaborative Archaeology with Apache Contexts." *Journal of Social Archaeology* 17, no. 1 (2017): 92–112.

Langhals, Brent T., Judee K. Burgoon, and Jay F. Nunamaker Jr. "Using Eye and Head Based Psychophysiological Cues to Enhance Screener Vigilance." Paper presented at the Credibility Assessment and Information Quality in Government and Business Symposium, Kauai, HI, January 4-7, 2011.

la paperson. *A Third University Is Possible*. Minneapolis: University of Minnesota Press, 2017.

Layton, Edwin. "Mirror-Image Twins: The Communities of Science and Technology in 19th-Century America." *Technology and Culture* 12, no. 4 (October 1971): 562-80.

Lee, Erika. "Enforcing the Borders: Chinese Exclusion along the U.S. Borders with Canada and Mexico, 1882-1924." *Journal of American History* 89, no. 1 (June 2002): 54-86.

Lenkersdorf, Carlos. "The Tojolabal Language and Their Social Sciences." *Journal of Multicultural Discourses* 1, no. 2 (2006): 97-114.

Lewis, Jason Edward, Noelani Arista, Archer Pechawis, and Suzanne Kite. "Making Kin with the Machines." *Journal of Design and Science*, July 16, 2018. https://doi.org/10.21428 /bfafd97b.

Leza, Christina. *Divided Peoples: Policy, Activism, and Indigenous Identities on the U.S.-Mexico Border*. Tucson: University of Arizona Press, 2019.

Lonetree, Amy. *Decolonizing Museums: Representing Native America in National and Tribal Museums*. Chapel Hill: University of North Carolina Press, 2012.

Lopez, Chairman Valentin. "Change on the Amah Mutsun Tribe: Local Steps to Take Now!" Paper presented at the Indigeneity and Climate Justice Conference, University of California, Santa Cruz, May 30, 2019.

Lopez-Maldonado, Julio Edgar. "Ethnohistory of the Stingless Bees *Melipona beecheii* (Hymenoptera: Meliponinae) in the Mayan Civilization, Decipherment of the Beekeeping Almanacs Part I in the 'Madrid Codex' and the Study of Their Behavioral Traits and Division of Labor." PhD diss., University of California, Davis, 2010.

Lugones, María. "Toward a Decolonial Feminism." *Hypatia* 25, no. 4 (Fall 2010): 742-59.

Lyon, David. *Surveillance Studies: An Overview*. Cambridge, UK: Polity, 2007.

Lyon, David. "Under My Skin: From Identification Papers to Body Surveillance." In *Documenting Individual Identity: The Development of State Practices in the Modern World*, edited by Jane Caplan and John Torpey, 291-310. Princeton, NJ: Princeton University Press, 2001.

Madrigal, Raquel A. "Immigration/Migration and Settler Colonialism: Doing Critical Ethnic Studies on the U.S.-Mexico Border." PhD diss., University of New Mexico, 2019.

Magnet, Shoshana A. *When Biometrics Fail: Gender, Race and the Technology of Identity*. Durham, NC: Duke University Press, 2011.

Maldonado-Torres, Nelson. "On the Coloniality of Being: Contributions to the Development of a Concept." *Cultural Studies* 21, no. 2-3 (2007): 240-70.

Mallery, Garrick. *Sign Language among North American Indians: Compared with That among Other Peoples and Deaf-Mutes*. Washington, DC: Smithsonian Institution-Bureau of Ethnology, 1881.

Marez, Curtis. *Farm Worker Futurism: Speculative Technologies of Resistance*. Minneapolis: University of Minnesota Press, 2016.

Martínez, María Elena. *Genealogical Fictions: Limpieza de Sangre, Religion, and Gender in Colonial Mexico.* Stanford, CA: Stanford University Press, 2008.

Marx, Karl. *Capital.* Edited by Frederick Engels. Translated by Samuel Moore and Edward Aveling. Moscow: Progress Publishers, 1954.

Masco, Joseph. *Theatre of Operations: National Security Affect from the Cold War to the War on Terror.* Durham, NC: Duke University Press, 2014.

Massey, Douglas S. "How Arizona Became Ground Zero in the War on Immigrants." In *Strange Neighbors: The Role of States in Immigration Policy*, edited by Carissa Byrne Hessick and Gabriel J. Chin, 40–60. New York: New York University Press, 2014.

McCoy, Alfred W. *Policing America's Empire: The United States, the Philippines, and the Rise of the Surveillance State.* Madison: University of Wisconsin Press, 2009.

McLachlan, Sean, and Charles River, eds. *The Apache Scouts: The History and Legacy of the Native Scouts Used during the Indian Wars.* CreateSpace, 2018.

Meeks, Eric V. *Border Citizens: The Making of Indians, Mexicans, and Anglos in Arizona.* Austin: University of Texas Press, 2007.

Mezzadra, Sandro, and Brett Neilson. *Border as Method, or, the Multiplication of Labor.* Durham, NC: Duke University Press, 2013.

Middleton, Robert D., Jr. "Privacy Impact Assessment Update for the Future Attribute Screening Technology (FAST)/Passive Methods for Precision Behavioral Screening." US Department of Homeland Security, Washington, DC, December 21, 2011. https://www.dhs.gov/sites/default/files/publications/privacy_pia_012a-s%26t_fast.pdf.

Miles, Nelson Appleton. *Personal Recollection and Observation of General Nelson A. Miles.* Chicago: Werner, 1896.

Miller, Andrea. "(Im)Material Terror: Incitement to Violence Discourse as Racializing Technology in the War on Terror." In *Life in the Age of Drone Warfare*, edited by Lisa Parks and Caren Kaplan, 112–33. Durham, NC: Duke University Press, 2017.

Miller, Todd. *Empire of Borders: The Expansion of the U.S. Border around the World.* New York: Verso, 2019.

Miller, Todd. *Storming the Wall: Climate Change, Migration, and Homeland Security.* San Francisco: City Lights Books, 2017.

Mitchell, Timothy. *Rule of Experts: Egypt, Techno-Politics, Modernity.* Berkeley: University of California Press, 2002.

Molina, Natalie. *Fit to Be Citizens? Public Health and Race in Los Angeles, 1879–1940.* Berkeley: University of California Press, 2006.

Mooney, James. "Sign Language." In *Handbook of American Indians North of Mexico*, edited by Frederick Webb Hodge, 567–68. Smithsonian Institution, Bureau of American Ethnology, Bulletin 30. Washington, DC: U.S. Government Printing Office, 1907.

Moreton-Robinson, Aileen. "Writing Off Indigenous Sovereignty: The Discourse of Secularity and Patriarchal White Sovereignty." In *Sovereign Subjects: Indigenous Sovereignty Matters*, edited by Aileen Moreton-Robinson, 86–104. Crows Nest, NSW, Australia: Allen and Unwin, 2007.

Morgan, Henry Lewis. *Ancient Society, or, Researches in the Lines of Human Progress from Savagery, through Barbarism to Civilization.* New York: Henry Holt, 1877.

Muñoz, José Esteban. *Cruising Utopia: The Then and There of Queer Futurity*. New York: New York University Press, 2009.

Munro, Campbell. "Mapping the Vertical Battlespace: Towards a Legal Cartography of Aerial Sovereignty." *London Review of International Law* 2, no. 2 (2014): 233–61.

Myer, Albert J. "A New Sign Language for Deaf-Mutes." PhD diss., University of Buffalo, 1851.

Nagpal, Radhika. "Taming the Swarm—Collective Artificial Intelligence." TEDx Talks, January 14, 2016, https://www.youtube.com/watch?v=LHgVRolzFJc.

Nakamura, Lisa. "Indigenous Circuits: Navajo Women and the Racialization of Early Electronic Manufacture." *American Quarterly* 66, no. 4 (December 2014): 919–41.

National Geographic Society. *Border Wars*. National Geographic Channel, 2010. Fifty-six episodes. https://www.natgeotv.com/za/shows/natgeo/border-wars

Nevins, Joseph. *Operation Gatekeeper: The Rise of the "Illegal Alien" and the Making of the U.S.-Mexico Boundary*. New York: Routledge, 2002.

Nott, Joseph Clark, and George R. Glidden. *Types of Mankind*. Philadelphia: Lippincott, Grambo, 1854.

O'Brien, Jean M. *Firsting and Lasting: Writing Indians out of Existence in New England*. Minneapolis: University of Minnesota Press, 2010.

O'Brien, Jean M. "Vanishing Indians in Nineteenth-Century New England: Local Historians' Erasure of Still-Present Indian Peoples." In *New Perspectives on Native North America: Cultures, Histories, and Representations*, edited by Sergi Kan and Pauline Strong Turner, 414–32. Lincoln: University of Nebraska Press, 2006.

Office of the Under Secretary of Defense for Acquisition, Technology, and Logistics. *Report of the Defense Science Board Task Force on Defense Biometrics*. Washington, DC: US Department of Defense, 2007.

O'Neil, Cathy. *Weapons of Math Destruction: How Big Data Increases Inequality and Threatens Democracy*. New York: Crown, 2016.

Organ Pipe Cactus National Monument. "Organ Pipe Cactus National Monument: Superintendent's 2010 Report on Natural Resource Vital Signs." Ajo, AZ: National Park Service, 2011. https://www.nps.gov/orpi/learn/nature/upload/orpi_vitalsigns2010.pdf, 4.

Painter, Fantasia. "Bordering the Nation: Land, Life, and the Law at the US-Mexico Border and on O'odham Jeved (land)." PhD. diss., University of California, Berkeley, 2021.

Palafox, Jose. "Opening Up Borderland Studies: A Review of U.S.-Mexico Border Militarization Discourse." *Social Justice* 27, no. 3 (2000): 56–72.

Park, Lisa Sun-Hee, and David Naguib Pellow. "Roots of Nativist Environmentalism in America's Eden." In *American Studies, Ecocriticism, and Citizenship: Thinking and Acting in the Local and Global Commons*, edited by Joni Adamson and Kimberly Ruffin, 175–89. New York: Routledge, 2012.

Parks, Lisa. "Drones, Infrared Imagery, and Body Heat," *International Journal of Communication* 8 (2014): 2518–21.

Parks, Lisa, and Caren Kaplan, eds. *Life in the Age of Drone Warfare*. Durham, NC: Duke University Press, 2017.

Peat, F. David. *Blackfoot Physics*. York Beach, ME: Red Wheel/Weiser, 2002.

Pegler-Gordon, Anna. *In Sight of America: Photography and the Development of U.S. Immigration Policy*. Berkeley: University of California Press, 2009.

Perez, Domino Renee. "New Tribalism and Chicana/o Indigeneity in the Work of Gloria Anzaldúa." In *The Oxford Handbook of Indigenous American Literature*, edited by James H. Cox and Daniel Heath Justice, 491–502. Oxford: Oxford University Press, 2014.

Pérez, Emma. *The Decolonial Imaginary: Writing Chicanas into History*. Bloomington: Indiana University Press, 1999.

Pérez, Laura. *Chicana Art: The Politics of Spiritual and Aesthetic Altarities*. Durham, NC: Duke University Press, 2007.

Postone, Moishe. *Time, Labor, and Social Domination*. Cambridge: Cambridge University Press, 1993.

Puar, Jasbir. "Precarity and the Politics of Nation: Settler States, Borders, Sovereignty." Paper presented at the Annual Conference of the National Women's Studies Association, Milwaukee, WI, November 13, 2015.

Puar, Jasbir. *The Right to Maim: Debility, Capacity, Disability*. Durham, NC: Duke University Press, 2017.

Pugliese, Joseph. *Biometrics: Bodies, Technologies, Biopolitics*. New York: Routledge, 2010.

Quezada-Euán, José Javier G., William De Jesús May-Itzá, and Jorge A. González-Acereto. "Meliponiculture in Mexico: Problems and Perspective for Development." *Bee World* 82, no. 4 (April 2015): 160–67.

Ramirez, Renya K. *Native Hubs: Culture, Community, and Belonging in Silicon Valley and Beyond*. Durham, NC: Duke University Press, 2007.

Ray, Sarah Jaquette. *The Ecological Other: Environmental Exclusion in American Culture*. Tucson: University of Arizona Press, 2013.

Redniss, Lauren. *Oak Flat: A Fight for Sacred Land in the American West*. New York: Random House, 2020.

Regnault, Félix. "Le Lange par gestes." *La Nature* 1324 (October 1898): 315–17.

Revels, Asa, and Janet Cummings. "The Impact of Drug Trafficking on American Indian Reservations with International Boundaries." *American Indian Quarterly* 38, no. 3 (2014): 287–318.

Riding In, James. "Six Pawnee Crania: Historical and Contemporary Issues Associated with the Massacre and Decapitation of Pawnee Indians in 1869." *American Indian Culture and Research Journal* 16, no. 2 (1992): 101–19.

Rifkin, Mark. *The Erotics of Sovereignty: Queer Native Writing in the Era of Self-Determination*. Minneapolis: University of Minnesota Press, 2012.

Rony, Fatimah Tobing. *The Third Eye: Race, Cinema, and Ethnographic Spectacle*. Durham, NC: Duke University Press, 1996.

Rose, Nikolas. *Powers of Freedom: Reframing Political Thought*. Cambridge: Cambridge University Press, 1999.

Rowe, Aimee Carrillo. "Settler Xicana: Postcolonial and Decolonial Reflections on Incommensurability." *Feminist Studies Journal* 43, no. 3 (2017): 525–36.

Roy, Deboleena. *Molecular Feminisms: Biology, Becomings, and Life in the Lab*. Seattle: University of Washington Press, 2018.

Sadowski-Smith, Claudia. "U.S. Border Ecologies, Environmental Criticism, and Transnational American Studies." In *American Studies, Ecocriticism, and Citizenship: Thinking and Acting in the Local and Global Commons*, edited by Joni Adamson and Kimberly Ruffin, 144–57. New York: Routledge, 2012.

Saldaña-Portillo, María Josefina. *Indian Given: Racial Geographies across Mexico and the United States*. Durham, NC: Duke University Press, 2016.

Saldaña-Portillo, María Josefina. *The Revolutionary Imagination in the Americas and the Age of Development*. Durham, NC: Duke University Press, 2003.

Salomón Johnson, Amrah Naomi. "Returning to Yuma: Regeneración and Futures of Autonomy." PhD diss., University of California, San Diego, 2019.

Sandoval, Chela. *Methodology of the Oppressed: Theory out of Bounds*. Minneapolis: University of Minnesota Press, 2000.

Santa Ana, Otto. *Brown Tide Rising: Metaphors of Latinos in Contemporary American Public Discourse*. Austin: University of Texas Press, 2002.

Saxton, Dean, Lucille Saxton, and Susie Enos. *Tohono O'odham/Pima to English: English to Tohono O'odham/Pima Dictionary*. Tucson: University of Arizona Press, 1998.

Schaeffer, Felicity Amaya. "Spirit Matters: Gloria Anzaldúa's Cosmic Becoming across Human/Nonhuman Borderlands." *Signs: Journal of Women in Culture and Society* 43, no. 4 (May 2018): 1005–29.

Schmidt, Louis Bernard. "Manifest Opportunity and the Gadsden Purchase." *Arizona and the West* 3, no. 3 (Autumn 1961): 245–64.

Schulze, Jeffrey M. *Are We Not Foreigners Here? Indigenous Nationalism in the U.S.- Mexico Borderlands*. Chapel Hill: University of North Carolina Press, 2018.

Shaw, William H. "Marx and Morgan." *History and Theory* 23, no. 2 (May 1984): 215–28.

Shorter, David Delgado. "Sexuality." In *The World of Indigenous North America*, edited by Robert Warrior, 487–505. New York: Routledge, 2015.

Shorter, David Delgado. "Spirituality." In *The Oxford Handbook of American Indian History*, edited by Frederick E. Hoxie, 433–52. Oxford: Oxford University Press, 2016.

Silko, Leslie Marmon. *The Almanac of the Dead: A Novel*. New York: Simon and Schuster, 1991.

Simmons, Kristen. "Settler Atmospherics." *Fieldsights*, November 20, 2017. https://culanth.org/fieldsights/settler-atmospherics.

Simpson, Audra. *Mohawk Interruptus: Political life across the Borders of Settler States*. Durham, NC: Duke University Press, 2014.

Simpson, Leanne Betasamosake. *As We Have Always Done: Indigenous Freedom through Radical Resistance*. Minneapolis: University of Minnesota Press, 2017.

Skelton, Ike. "America's Frontier Wars: Lessons for Asymmetric Conflicts." *Military Review* 81, no. 5 (September–October 2001): 22–27.

Smith, Andrea. *Conquest: Sexual Violence and American Indian Genocide*. Cambridge, MA: South End, 2005.

Smith, Andrea. "Not-Seeing: State Surveillance, Settler Colonialism, and Gender Violence." In *Feminist Surveillance Studies*, edited by Rachel E. Dubrofsky and Shoshana Amielle Magnet, 21–38. Durham, NC: Duke University Press, 2015.

Smith, Cornelius C., Jr. *Fort Huachuca: The History of a Frontier Post*. Fort Huachuca, Arizona: US Government Printing Office, 1977.

Solnit, Rebecca. *Savage Dreams: A Journey into the Hidden Wars of the American West.* Berkeley: University of California Press, 1994.

Sotelo Santos, Laura Elena, and Carlos Alvarez Asomoza. "The Maya Universe in a Pollen Pot: Native Stingless Bees in Pre-Columbian Maya Art." In *Pot-Pollen in Stingless Bee Melittology*, edited by Patricia Vit, Silvia R. M. Pedro, and David W. Roubik, 299–309. Cham, Switzerland: Springer, 2018.

Stern, Alexandra Minna. *Eugenic Nation: Faults and Frontiers of Better Breeding in Modern America.* Berkeley: University of California Press, 2005.

Tallbear, Kim. "Beyond the Life/Not-Life Binary: A Feminist-Indigenous Reading of Cryopreservation, Interspecies Thinking, and the New Materialisms." In *Cryopolitics: Frozen Life in a Melting World*, edited by Joanna Radin and Emma Kowal, 179–202. Cambridge, MA: MIT Press, 2017.

Tallbear, Kim. "Making Love and Relations beyond Settler Sex and Family." In *Making Kin, Not Populations*, edited by Adele Clarke and Donna Haraway, 145–57. Chicago: Prickly Paradigm, 2018.

Tallbear, Kim. *Native American DNA: Tribal Belonging and the False Promise of Genetic Science.* Minneapolis: University of Minnesota Press, 2013.

Tambiah, Stanley. *Magic, Science, Religion and the Scope of Rationality.* Cambridge: Cambridge University Press, 1990.

Terry, Jennifer. *Attachments to War: Biomedical Logics and Violence in Twenty-First-Century America.* Durham, NC: Duke University Press, 2017.

The Discovery Channel. *Border Live.* Stamford, CT: Lucky 8, 2018. Three episodes. https://www.discoveryplus.com/show/border-live.

Thrapp, Dan L. "The Indian Scouts, with Special Attention to the: Evolution, Use, and Effectiveness of the Apache Indian Scouts." In *Military History of the Spanish American Southwest: A Seminar*, n.p. Fort Huachuca, AZ: US Army Commo. Cmd., 1976.

Todd, Zoe. "Refracting the State through Human-Fish Relations: Fishing, Indigenous Legal Orders and Colonialism in North/Western Canada." *Decolonization: Indigeneity, Education and Society* 7, no. 1 (2018): 60–75.

Trope, Jack F., and Walter R. Echo-Hawk. "The Native American Graves Protection and Repatriation Act: Background and Legislative History." *Arizona State Legislature* 24 (Spring 1992): 35–77.

Tsing, Anna. "Empowering Nature, or: Some Gleanings in Bee Culture." In *Naturalizing Power: Essays in Feminist Cultural Analysis*, edited by Sylvia Yanagisako and Carol Delaney, 113–44. New York: Routledge, 1995.

Tuck, Eve, and K. Wayne Yang. "Decolonization Is Not a Metaphor." *Decolonization: Indigeneity, Education and Society* 1, no. 1 (2012): 1–40.

Urbanski, Claire. "On Sacred and Stolen Lands: Desecration and Spiritual Violence as United States Settler Colonial Dispossession." PhD diss., University of California, Santa Cruz, forthcoming 2022.

US Department of Homeland Security. *Environmental Assessment for Integrated Towers on the Tohono O'odham Nation in the Ajo and Casa Grande Stations' Areas of Responsibility.* US Border Patrol Tucson Sector, Arizona US Customs and Border Protection, Department of Homeland Security, Washington, DC, March 2017. https://www.cbp.gov

/sites/default/files/assets/documents/2017-Apr/TON%20IFT%20FINAL%20EA%20
FONSI%202017%2003%20Part%20I.pdf.

US Department of the Interior, National Biological Service. "Organ Pipe Cactus
National Monument, Ecological Monitoring Program, Monitoring Protocol
Manual." Special Report No. 11. Tucson: National Biological Service, Cooperative Park
Studies Unit, University of Arizona, 1995. https://babel.hathitrust.org/cgi/pt?id=mdp
.39015038419274&view=1up&seq=7.

US Department of Justice (Sean B. Hoar, s.b.). "Identity Theft: The Crime of the New
Millennium." *USA Bulletin* 49, no. 2 (2001): 1–15.

Varela, Francisco. *Ethical Know-How: Action, Wisdom and Cognition*. Stanford, CA: Stanford
University Press, 1999.

Villanueva-Gutiérrez, Rogel, David W. Roubik and Wilberto Colli-ucán. "Extinction of
Melipona beecheii and Traditional Beekeeping in the Yucatán Peninsula." *Bee World* 86,
no. 2 (June 2005): 35–41.

Virilio, Paul. *War and Cinema: The Logistics of Perception*. Translated by Patrick Camiller.
London: Verso, 1989.

Wagner, Arthur L. *Organization and Tactics*. 1894. Kansas City, MO: Franklin Hudson,
1906.

Wagner, Arthur L. *The Service of Security and Information*. Kansas City, MO: Franklin
Hudson, 1893.

Wallace, Edward S. *The Great Reconnaissance: Soldiers, Artists, and Scientists on the Frontier,
1841–1861*. Boston: Little, Brown, 1955.

Walsh, James. "Border Theatre and Security Spectacles: Surveillance, Mobility and
Reality-Based Television." *Crime Media Culture* 11, no. 2 (2015): 201–21.

Watts, Vanessa. "Indigenous Place-Thought and Agency amongst Humans and Non
Humans (First Woman and Sky Woman Go on a European World Tour!)." *Decolonization: Indigeneity, Education and Society* 2, no. 1 (2013): 20–34.

Weaver, Nervin, and Elizabeth C. Weaver. "Beekeeping with the Stingless Bee *Melipona
beecheii*, by the Yucatecan Maya." *Bee World* 62, no. 1 (July 2015): 7–19.

Webb, Walter Prescott. *The Great Plains*. Lincoln: University of Nebraska Press, 1981.

Whyte, Kyle Powys. "On the Role of Traditional Ecological Knowledge as a Collaborative Concept: A Philosophical Study." *Ecological Process* 2 (2013): article 7. https://doi
.org/10.1186/2192-1709-2-7.

Willey, Angela. "Biopossibility: A Queer Feminist Materialist Science Studies Manifesto,
with Special Reference to the Question of Monogamous Behavior." *Signs: Journal of
Women in Culture and Society* 14, no. 3 (2016): 553–77.

Willis-Grider. Marilyn. "Cross-Cultural Competence." *MIPB* (*Military Intelligence Professional Bulletin*) (2011): 2–9.

Wolfe, Patrick. "After the Frontier: Separation and Absorption in US Indian Policy." *Settler Colonial Studies* 1, no. 1 (2011): 13–51.

Wolfe, Patrick. "Settler Colonialism and the Elimination of the Native." *Journal of Genocide Research* 8, no. 4 (2006): 347–409.

Yandell, Kay. *Telegraphies: Indigeneity, Identity, and Nation in America's Nineteenth-Century
Virtual Realm*. New York: Oxford University Press, 2019.

Yellowman, Connie Hart. "'Naevahooohtseme'—We Are Going Back Home: The Cheyenne Repatriation of Human Remains—a Woman's Perspective." *St. Thomas Law Review* 9, no. 1, (1996): 103–16.

Zaytoun, Kelli. "'Now Let Us Shift' the Subject: Tracing the Path and Posthumanist Implications of La Naguala/The Shapeshifter in the Works of Gloria Anzaldúa." *MELUS: Multi-Ethnic Literature of the U.S.*, 40, no. 4 (2015): 69–88.

Zepeda, Susy. "Queer Xicana Indígena Cultural Production." *Decolonization: Indigeneity, Education and Society* 3, no. 1 (2014): 119–41.

Zuboff, Shoshana. *The Age of Surveillance Capitalism: The Fight for a Human Future at the New Frontier of Power*. New York: Hachette, 2019.

beekeeping, 21, 26, 105; death and, 179n83; erotic relationality of, 124–25, *125, 126,* 130; revival of *Melipona,* 121–22; sacred science of, 137, 138; shared destinies of, 135–36; women caretakers and collectives, 122–23, 136

bees: for bomb detection, 111; cosmologies, 131–32; in DARPA's branding strategy, 109, *110;* debate over consciousness of, 114; decline and comeback, 112, 121–22, 176n11, 176n25, 178n53; drones, 104–5; hierarchy, 10–11; intelligence, 106–7, 115–17; "killer," 118; as models for social-political transformation, 117; ownership rights to, 9; pesticides and, 178n57; robotic pollinating, 106, 112; scouts, 120–21; sensitivity of, 133–34, 179n83; surveillance towers and, 71

behavioral pattern recognition, 92–93, 97–99

belonging, 10, 25, 48, 55; citizenship and, 58, 59; with land, 7, 79, 130, 141

Benton, Lauren, 56

Bianchini, Marcello Levi, 87–88

big data, 5, 97

binoculars, 25, 36, 38

biolegitimacy, 88–89, 171n31

biomass, 89, 120, 171n36

biometrics, 84, 85, 87–89, 99, 169n1; crimes and national databases and, 101–2. *See also* AVATAR (Automated Virtual Agent for Truth Assessment in Real-Time)

biorobotics, 112, *113,* 119, 120

biosecurity-industrial complex, 83, 85–86

black holes, 55, 66, 78, 90, 163n1

Blackness, 17

bodies: biometrics and, 84, 87; bones, 142; cellular biology, 113; control of, 5, 101; land and, 2, 4, 15, 16, 35, 81, 129–30; physiological movements, 81, 84, 93–94, 99; race/racialization and, 17, 88–90, 102, 174n84; scanning and data collection, 25–26, 72, 81–82, *82,* 101; seeing/unseeing, 9, 24; signs of deceit, 91–93; targeting of, 100. *See also* emotional expressions

border checkpoints, 82, 83; Israeli, 75, 98; Nogales (AZ), 81, 95; Singapore-Malaysia, 95; Tohono O'odham reservation, 57, 58, 59, 66

Borderlands/La Frontera (Anzaldúa), x–xi, 8, 129, 151–52

Border Patrol, 33, 50, 96, 155n14; all-Native American (Shadow Wolves), 6, 42, 66, 76–78; automated, 81; funding increases, 85–86; national park rangers as, 145; O'odham sovereignty and, 56, 57, 58, 60–61, 68–69, 72; US-Canadian, 78–79

border-security companies, 96

border walls: digital, 99–100, 102; Israeli, 97; Trump's construction, 10, 64–65, 67, 86, 139–41, 144–45, 147, 151–52, 180nn1–2

Brandt, Elizabeth, 19

Breault, Robert, 169n2

Brooks, Rodney, 119

Brown, Wally, 39

Browne, Simone, 17, 88

Bruyneel, Kevin, 155n16

Büchner, Ludwig, 107, 116–17, 177n34

buffalo, 71

Burgoon, Judee, 91, 92–93

burial grounds, 27, 83, 150, 158n59, 181n4; legal protection, 142; Tohono O'odham, 18, 26, 139, 144, 147–48, 180n2

Butler, Judith, 155n20

Byrd, Jodi, 3, 9, 154n5, 164n21

Cáceres, Berta, 124

Cajete, Gregory, 22, 41

camouflage, 41, 52

Canada-US border, 56, 58, 79, *82*

Castellanos, M. Bianet, 136

Castillo, Ana, x, 153n2

Centers for Disease Control and Prevention (CDC), 174n92

ceremonies, 36, 57, 72, 73, 124, 131, 150

checkpoints. *See* border checkpoints

Chicanos, x, 153n3

Chiricahua Apache, 3, 4, 6, 19, 48, 150, 160n13

Choctaw, 6, 159n3

Chow, Rey, 100

citizenship, 7, 58–59, 62, 80, 170n6; false claims to, 85; land title and, 63–64; O'odham, 64, 69, 70, 75, 167n77

Civil War, 32, 40, 160n10, 162n68

climate change, 69–70, 145

Code of Indian Offenses (1883), 158n65

code (or wind) talkers, 6, 50, 160n10

Cold War, 14, 93, 109

colonial imaginary, 6, 46, 54, 59

commemorations, 156n33, 160n10; statues, 29, 30, 31, 35–36, 159n1

communication technologies, 39–42, 78; lie detectors, 93–94; telegraph lines, 12–13, 53

concealment, 45–46, 53

consciousness: animal, 114, 119, 177n34; expanding human, 128–29

conservation, 26, 123, 140–42, 145–46, 150

Cortés, Don José, 44

Cortés, Hernán, 161n42

cosmologies, 8, 18, 23, 34, 127, 153n2; Anzaldúa's Mesoamerican, x–xi, 128–29, 156n22; bee, 125–26, *126, 127,* 131–32; of footprints, 6; Tohono O'odham, 57, 69. *See also* sacredsciences

COVID-19 pandemic, 102–3

criminalization, 56, 75, 88, 158n65, 168n79; detecting intent, 93, 94; of migrants/immigrants, 80, 85–86, 140; national databases for, 101–2

Cristiano, Don, 133

Crook, General George, 48, 50–51

Darwin, Charles, 26, 40, 92, 122, 171n46; bee studies, 107, 115–17, 177nn38–40; scientific method, 177n35

healers, 41, 122, 130

helicopters, 72–73

heliograph stations, 51–52, 162n68

heroes, Native Americans as, 32–33, 160n10

hierarchy, 47, 107, 115, 119

Him'dag, meaning, 57, 58, 68, 69

Hobart, Hi'ilei Julia, 18

Hochman, Brian, 12, 39, 160n18

Hohokam, 27, 139

honeybees (*Apis mellifera*), 111, 112, 124; European, 115, 117, 118, 122, 177n38; waggle dance of, 120–21

hot spots, 66, 89–90

Hu, Tung-Hui, 12

Huachuca Illustrated, 36, 161n19

Huachuca mountain range, 4, 6, 20, 33–34, 51, 150

Hualapai Tribe, 74

human-animal relations, 92, 103, 107, 175n11; deities and, 125–26; methods of becoming, 108–9, 133, 136; physical transformations, 41, 49–50; sensual or spiritual connections, 131–34. *See also* beekeeping; more-than-human world

human intelligence, 17, 31, 47, 89, 119; bee intelligence and, 107, 116–17; technology and evolution and, 10–12

humble bees, 115–16

hunting practices, 41, 49, 50, 72, 77

iconography, 36, 38

identity, 5, 130, 170n6; biometric technologies and, 81, 86, 101; Maya, 123, 136; mestizo/mestiza, 155n19, 155n21; theft, 84, 85

Illegal Immigrant Reform and Immigrant Responsibility Act (IIRIRA), 85, 86, 174n90

Indian, term usage, 159n2

Indian Citizenship Act (1924), 63–64, 160n12

Indianness, 12, 54. *See also* Indigeneity

Indian scouts, 8, 25; Alamo, 50; Alexa as, 12; communication practices of, 38–39; hired by the US Army, 4, 31, 44–47, 50, 162n49; Sioux, 160n17; statue and photo of, 29, 30, 31, 35–36, 38; tracking Geronimo, 51. *See also* Shadow Wolves

Indian Wars, 15, 34–35, 53, 54, 139, 155n14, 160n17; end of, 4, 36, 42; lessons from, 52; term usage, 155n12; timing of, 155n11

Indigeneity, 3, 54, 60, 155n13, 155n21; Americanness and, 5; Anzaldúa on, 126, 129, 151; September 11 attacks and, 14; technological innovations and, 15, 32

Indigenous knowledges, 2–5, 144, 151, 153n2, 158n59; Anzaldúa and Mesoamerican, x, xi, 127, 129, 130; bees and, 106; land or place-based, 19–20, 35, 41, 140, 158n70; Nishnaabeg, 20–21; tracking skills, 45, 77–78; traditional ecological knowledge (TEK), 8, 21–22, 158n73; Western science and, 18, 34–35, 158n59. *See also* Nativision; sacredsciences

Indigenous studies, 2, 3, 21, 59, 129, 155n19; ethnological studies, 40

individualism, 5, 47, 48, 171n36

infrared imaging, 90, 100

insects, 106–7, 116, 120, 175n11; termites, 118–19. *See also* bees

instinct, 115, 116

intellectual property rights, 94

intelligence. *See* human intelligence; military intelligence; swarm intelligence

interbeing, 136

invisibility, 45, 48, 52, 154n7

Iraq, 4, 42, 52–53, 175n3

Iroquois, 10, 56, 58, 163n8

Israeli bordering technologies: checkpoints, 75; drones, 71, 97; security companies, 96; Suspect Detection Systems (SDS), 97–98

Israeli Homeland Security, 97–98

Jose, Verlon M., 58, 67–69, 80

Kao, John, 13–14, 157n41

Kaplan, Caren, 14

Kickapoo, 2, 56, 163n8

Kilobots, 113, 119–20

Kimmerer, Robin, 22

kin, making, 13, 109, 176n14

Koolel-Kab, 123, 136

Kosek, Jake, 112, 175n11

labor, 109, 114, 118, 125, 143–44; automated, 119–20; of beekeeping, 125, 131, 133, 137; creative, 11–12, 120; markets, 70, 97, 118; ritual and, 128, 129

LaDuke, Winona, 19, 32, 160n10, 160n12

Laluk, Nicholas, 19

land claims, xi, 3, 7, 38, 151, 154n7, 155n21; federal, 141, 144, 181n8, 181n12; legal doctrines, 9; military's role in, 160n10

land grabs, 64, 65, 124, 160n10

land leasing, 63, 64, 71, 165n33

la paperson, 9, 156nn25–26

Latinx, 7, 85, 130; migrants, 25, 27, 102; studies, x, 2, 3, 59; term usage, 153n3

lawless spaces, 55, 56, 66, 143–44

Layton, Edwin, 160n7

Lenkersdorf, Carlos, 136

Levy, Steven, 99

lie detectors, 93–94, 97

Liston, Stanley, 76

Little Bear, Leroy, 158n57

Lopez, Val, 24

Los Alamos National Laboratory, 106, 111

Luckey, Palmer, 99–100

Lyon, David, 157n55

Madrid Codex, 124, *125*, 132–33

Mallery, Garrick, 40–41, 161n32

maps/mapping, 6–7, 34, 40, 43, 72, 147; Indigenous guides, 44, 61, 164n27; US-Mexico border, 61–62, 67